T0135524

Domain-Specific Development of Event Condition Action Policies

PhD Thesis

Software Methodologies for Distributed Systems
Faculty of Applied Computer Science
University of Augsburg, Germany

Raphael Romeikat

Supervisor:	Prof. Dr. Bernhard Bauer
Advisor:	Prof. Dr. Theo Ungerer
Date of defense:	15th October 2012

Bibliographic information published by the Deutsche Nationalbibliothek

The Deutsche Nationalbibliothek lists this publication in the Deutsche Nationalbibliografie; detailed bibliographic data are available in the Internet at http://dnb.d-nb.de .

ISBN 978-3-8325-3761-6

Logos Verlag Berlin GmbH
Comeniushof, Gubener Str. 47,
10243 Berlin
Tel.: +49 (0)30 42 85 10 90
Fax: +49 (0)30 42 85 10 92
INTERNET: http://www.logos-verlag.de

Abstract

Information Technology (IT) systems have a high impact on organizations and businesses today and are a key to globalization. Evolution of technology has changed the nature of these systems during the last decades. Growing computer performance enables more complex hardware and software technology and allows for more complex system architectures. This trend complicates the development and management of IT systems.

Evolution of complexity calls for changes in the way IT systems are built and managed. As a consequence, a paradigm shift towards Model-Driven Engineering (MDE) approaches can be observed. MDE techniques help to reduce complexity at design time by raising the level of abstraction and automation with models and transformations that generate major parts of source code. Policy-Based Management (PBM) addresses the abstraction and automation of a system at runtime. It enables modifications of system behavior without considering technical details or changing source code. Policies define which actions must or must not be performed according to the particular situation. A well-known application domain is network management, where policies are widely used for automated configuration processes.

In this thesis, we present an innovative approach to the domain-specific development of Event Condition Action (ECA) policies. Our approach separates domain and policy aspects into different models and thus facilitates the collaboration of different expert groups. It initially represents the models at a high level of abstraction and generates refined, lower-level policies in an automated way. Finally, the refined models can be transformed into executable code. In our approach, models do not only serve documentation purposes, but represent essential artifacts of the policy development process. Their refinement into a machine-executable representation allows to change system behavior by changing the high-level models even at runtime as high-level changes are automatically reflected in the low-level representation of the models. We use a relational algebra to formally specify and validate the models at each layer and to define the formal semantics of the refinement. We also prove that the refinement preserves the static semantics of the models. Finally, we demonstrate and evaluate our approach with two real-world case studies that deal with the management of mobile networks and the calculation of bonus payments for employees.

Acknowledgments

I would like to thank all people that supported me in writing my thesis. First of all, I am deeply grateful to my supervisor Prof. Dr. Bernhard Bauer for providing me with the opportunity to conduct my research and for his excellent mentoring and valuable comments. His mindful support represented an important source of inspiration. I also thank Prof. Dr. Theo Ungerer, who accepted to be the advisor of my thesis.

Special thanks go to Nokia Siemens Networks Research for their continuous funding and support of my research. The joint work with Dr. Henning Sanneck, Christoph Schmelz, Tobias Bandh, and further colleagues provided great motivation and very interesting use cases for my research. Nothing has challenged me more than our outstanding Currywurst in Berlin.

I thank my colleagues at the Software Methodologies for Distributed Systems Lab for creating a friendly and cheerful atmosphere that I enjoyed a lot. Scientific discussions about challenging questions and innovative solutions definitely contributed to my thesis. My thanks include all students who assisted me in my research and helped me with their implementations to save time.

Furthermore, I thank all other people who directly or indirectly supported me in writing my thesis. All informal support by my friends and by researchers who I met at international conferences had a very positive influence on myself and my research.

Last but not least, I especially thank my parents Sia and Ralf for their encouragement and for providing me with an excellent environment to perform my studies, and my sister Raqui for her time and her friendship. The most important thanks go to my dearest Elli for her endless love, understanding, and support in all matters. She is the most important person for me and gives an incredible meaning to my life.

Contents

1

Introduction

Information Technology (IT) has a high impact on organizations and businesses today and is a key to globalization. In 2010, businesses in the United States spent $263.1 billion on Information and Communication Technology (ICT) equipment and computer software [1]. Various reasons exist for these investments. Businesses intend to reduce their costs, increase their sales, and raise their productivity.

Today's ICT systems are the result of a long lasting development in computing technology. Growing computer performance has enabled more complex hardware and software technology and allows for complex system architectures today. Increasing complexity of information systems complicates their development and management and further progress in technology will lead to even more complex systems.

As a response to this evolution, different approaches and techniques have been established. One approach to cope with complexity is Policy-Based Management (PBM). A key point of PBM is the reduction of system complexity by automation. Policies represent reaction rules for particular situations and can be enforced automatically at runtime. Whereas the aspect of automation has been well investigated, abstraction in policy development remains a widely open issue. Little attention has been paid on the specification of policies at an abstract level and on their refinement into a technical representation. Due to the lack of appropriate solutions, we propose an innovative approach to the development of policies. In our approach, we use models to specify policies at different abstraction layers, and we refine them with model transformations in an automated way.

In the following, we summarize in section 1.1 how complexity has evolved in the recent decades. In section 1.2, we then analyze problems that arise from increasing complexity and derive appropriate objectives. In section 1.3, we summarize our approach to achieve these objectives, and we provide our contributions that realize the approach. We list our relevant publications in section 1.4. Finally, we outline the structure of this thesis and provide an overview of the chapters in section 1.5.

1.1 Evolution of Complexity

During the last decades, evolution of technology has changed the nature of ICT systems. In 1965, Moore observed that the number of transistors on an integrated circuit had doubled every year since the invention of the integrated circuit, and that cost had decreased at a similar rate, as shown in figure 1.1. He also predicted that this trend would continue for the following ten years. In 1975, Moore revisited his prediction and changed it to doubling every couple of years in future [2]. The prediction often quoted as Moore's Law was that computer performance would double every 18 months, although Moore has never said that explicitly. Other measures of technology have developed at exponential rates similar to Moore's Law. Examples are increasing power consumption of compute nodes [3] and decreasing hard disk storage cost per unit of information [4], also known as Kryder's Law.

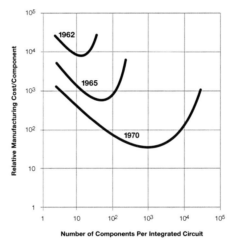

Figure 1.1: Moore's Law [5]

Since those days, evolution of hardware technology has also enabled evolution of software technology. Various programming models have evolved, starting from assembly languages over structured and component-oriented programming to compiled high-level languages and Service-Oriented Architectures (SOAs). Software usually undergoes a growth in size and complexity during its lifetime due to the addition of functionality. The discipline of software engineering uses development processes and structuring techniques such as software patterns and models in order to reduce complexity and increase productivity. In addition to software complexity, there has also been a substantial growth in system complexity since Moore. Various reasons exist for that, such as the variety of technologies deployed or ubiquitous system access on 24 hours a day and seven days a week. Often, systems are used by a huge and increasing number of users, and the need for scalability and robustness results in complex and highly distributed system architectures that avoid bottlenecks and single points of failure.

Predictions about the validity of Moore's Law for the future vary. Some experts give physical reasons in terms of smallest possible transistor size and device density and argue that Moore's Law will only hold for another one or two decades [6,7]. Others consider next generations of supercomputers and extend Moore's Law from semiconductor integrated circuits to other forms of technology and predict a much longer validity [8,9]. In either case, progress in technology will lead to more complex systems as the human thirst for computation will never slow down. Quite the contrary, development and maintenance of future systems will require more staff, resulting in more effort and costs than today. At the same time, there will be a reduction in both skill and productivity per person as the staff size increases [10]. Increasing complexity might even become a limiting factor to the further evolution of ICT systems. IBM considers this trend the main obstacle to further progress in the IT industry and calls it the complexity crisis [11]. Other experts call for a revolution to develop efficient and flexible ICT infrastructures [12]. Thus, new ways must be found to develop and manage modern systems. This research area currently receives much attention and is considered a hot topic in industry and academia.

As a consequence of increasing complexity, a paradigm shift towards Model-Driven Software Development (MDSD) [13] can already be observed. As a specialization of Model-Driven Engineering (MDE) [14], MDSD raises the level of abstraction and reduces complexity for software and system developers. It moves the focus from a code-centric to a model-centric point of view and allows for propagation of changes at the functional level down to the technical level of abstraction. At the same time, automated model transformations enable to automatically generate major parts of source code and thus reduce time for development and raise productivity. Besides MDSD, other approaches also realize MDE principles, such as Model Driven Architecture (MDA) [15], software factories [16], or domain-oriented programming [17].

Policy-Based Management (PBM) is another prominent approach to address the challenge of complexity. The idea behind PBM is controlling and managing a system at a high level of abstraction to relieve administrators from recurring tasks. Policies are considered an appropriate means for modifying the behavior of a system without changing source code and considering technical details of the system [18]. A system can continuously be adjusted to externally imposed constraints by changing the determining policies. Policies specify which actions must or must not be performed according to the particular situation. A situation is often characterized by a set of conditions that are evaluated on the occurrence of certain events. The respective actions are executed automatically as soon as the corresponding conditions are met. Such Event Condition Action (ECA) policies often exist at different levels of abstraction, which are considered different layers of a policy continuum [19].

IBM was the first well-known IT manufacturer to identify and address the challenge of complexity. They founded the Autonomic Computing Initiative (ACI) and raised the vision of computing systems that manage themselves according to high-level objectives defined by administrators. An important benefit is the reduction of time and cost for system administration [20, 21]. Autonomic Computing Systems (ACSs) must be capable of the four self-x properties: self-configuration, self-healing, self-optimization, and self-protection. The ACI vision was inspired by the human autonomic nervous system, which independently adjusts low-level vital functions and allows the brain to deal with more important high-level tasks. Other IT manufacturers also raised respective initiatives, such as the adaptive enterprise strategy by Hewlett-Packard [22], the dynamic systems initiative by Microsoft [23], and the N1 initiative by Sun [24].

The Organic Computing (OC) initiative [25] was a priority program funded by the German Research Foundation (DFG) and addressed growing complexity as well. It covered the objectives of the ACI and assumed that we would be surrounded by a lot of intelligent systems in future, which would be aware of their environment, communicate freely, and organize themselves in order to act in an autonomous way. Such systems were called Organic Computer Systems (OCSs) due to their life-like nature and their organic properties with respect to their architecture and functionality. OCSs are equipped with sensors and actuators and built of intelligent components and subsystems that do not only apply algorithms to divide labor and fulfill a fixed task. Instead, they evolve over time and dynamically adapt themselves to the conditions of their environment in order to meet upcoming challenges. However, the autonomy of OCSs also involves a problem of controllability. For this reason, the OC initiative aimed at developing generic control mechanisms in order to construct systems as robust, safe, and trustworthy as possible.

1.2 Problems and Objectives

In this section, we identify major problems with regard to the development and operation of complex systems, and we analyze the causes of their existence. To emphasize their importance, we mention relevant effects and challenges that arise from them. We also derive objectives from these problems and point out the importance of addressing and solving them. These objectives represent the main focus of this thesis and give the impression of an ideal situation in which the identified problems are no more present.

Software has an inherent level of complexity by nature as it is intended to solve problems, and problems have an inherent level of complexity. Moreover, solving a problem with software tends to add additional complexity beyond the problem itself. As software addresses problems of increasing complexity today, projects become larger with more and more developers involved. Increasing problem complexity leads to increasing software complexity, which in turn provokes increasing system complexity. Furthermore, the requirements on modern large-scale systems more and more grow. The need for scalability, availability, security, integration, and robustness results in even more complex and distributed system architectures. A lot of problems call for the application of various techniques, and this further increases complexity. Recent IT failure research revealed that 62 percent of IT projects are not completed on time and 49 percent suffer from budget overruns [26].

Development of complex systems demands for experts from various fields such as business experts, application experts, and technical experts to collaborate with each other during the development. Software development processes often partition complexity by passing through numerous iterations, specifying further details and realizing further aspects of a system step by step. Although software quality still performs at an unsatisfactory level, iterative development and enhanced project management have already led to an improvement. According to the CHAOS Report, 35 percent of software projects started in 2006 were successful [27], i.e. they were completed on time, on budget, and properly met user requirements, compared to only 16.2 percent according to the groundbreaking report in 1994 [28].

Operation of complex systems is another issue. Today, systems connect more and more physical and logical devices such as databases, applications, user interfaces, web servers, proxies, and firewalls. General challenges of large-scale distributed systems are the dynamics of their environment and the complexity of their operation. These are often driven by heterogeneous environments as devices run on various hardware configurations. Thus, administrators must comprehend different operational paradigms and complex information flows in order to diagnose a problem and find a solution.

Increasing complexity of IT systems makes their operation a time-consuming, expensive, and error-prone task. As a result, this task can hardly be performed by administrators any more. About 40 percent of computer system outages are caused by human errors [29], but do not happen because of lacking training or capabilities. Instead, the complexity of the systems exceeds human capabilities. Administrators cannot keep up with such complexity, particularly in stressful situations when decisions must be made within a very short period of time [30]. It is therefore not surprising that, according to estimations, one third to one half of a company's total IT budget is spent on preparing against or recovering from crashes [31].

Systems usually evolve over the years, and they tend to grow during their lifetime due to modification or addition of functionality. Taking the example of mobile networks, the addition of new Radio Access Technologys (RATs) has led to the coexistance of GSM, UMTS, and LTE today [32]. The laws of software evolution implicate a continual need for adaptation, resulting in continuing change, continuing growth, and increasing complexity [33]. Several empirical studies were undertaken to analyze commercial and open-source software and to describe such evolution of complexity. As a result, various mathematical models were proposed, describing polynomial functions to measure growth of complexity over time [34–36].

The identified challenges in the development and operation of complex systems lead to the following problem and objective definitions.

Problems

A lot of human effort is necessary for the development and operation of complex systems, besides intensive collaboration of different expert groups, which often proves difficult and inefficient.

1. Different experts and expert groups use different terminology to describe a system and the underlying domain due to their different background, vocabulary, and expertise.

2. Complex systems do not only consist of static components, but are often based on distributed and dynamic architectures, and this makes it difficult for developers to describe the behavior of the overall system.

3. Collaboration of application experts and technical experts to align high-level descriptions of system behavior with their low-level implementation is a time-consuming and error-prone task.

4. Systems in a dynamic environment are often expected to undergo frequent changes, demanding for frequent adaption of system behavior at runtime and involving a lot of human effort.

Objectives

The development and operation of complex systems should be facilitated and automated, and a common understanding of the domain and common means to describe the system behavior should be available.

1. Each expert group should agree on their individual understanding of the domain; for a better understanding between expert groups, all of them should use the same means to encapsulate domain knowledge.

2. The behavior of a dynamic system should be described with decoupled rules to enable frequent adaption; each expert group should specify these rules with their individual focus, but with the same means.

3. Higher-level descriptions of system behavior should be refined in an automated way for further processing; lower-level descriptions should be generated as input for the technical experts.

4. It should be possible to change the behavior of a system at runtime and in a flexible way, in particular without manually changing the implementation of the system.

1.3 Approach and Contributions

In the previous section, we identified problems concerning the development and operation of complex systems, and we derived appropriate objectives. Now, we present an approach that solves the identified problems and allows to achieve the derived objectives. Furthermore, we summarize our contributions, which represent the building blocks to realize the presented approach. They represent the results of this thesis and consist of both theoretical concepts and applicable artifacts. We also describe the techniques we use in the contributions.

As a basic principle, we propose the usage of policies to reduce complexity by abstraction and automation. ECA policies are an appropriate means to specify the reactive behavior of complex systems, i.e. how to behave in particular situations. They represent reaction rules that are independent of technical aspects and therefore raise the level of abstraction. The usage of policies enables modifications of system behavior without changing the underlying implementation, so that the system can continuously be adjusted to changing requirements by changing the determining policies [37].

Apart from abstraction, policies enable a high degree of automation. They encapsulate decision logic by describing which actions must be executed under which circumstances. Such circumstances are specified by the occurrence of events and the evaluation of conditions that define whether the policy is applicable or not. Whenever an ECA policy is triggered by an event, it is automatically enforced by the policy engine. For this purpose, the engine evaluates the policy condition and if the condition is met, enforces the execution of the policy actions.

In order to enable an effective usage of policies for the development and operation of complex systems, we developed an approach that supports the developer in the specification of policies and automates the refinement of the specified policies as far as possible. Figure 1.2 provides the big picture of our approach. The degree of automation is illustrated with colors: manual steps are highlighted in orange and automated steps in green. The contributions of our approach are highlighted in gray.

To realize our approach, we make use of techniques from MDE. Models allow to simplify the view of the often complex reality by separating different concerns from each other. Each model is an instance of a respective metamodel. The metamodel defines the abstract syntax of the model, i.e. its structure, independently of any concrete notation. Common entities in the metamodels define how to put model instances together, as indicated in figure 1.2 with connecting lines in gray. We perform the separation of concerns in two dimensions, and this separation results in different models at different layers, as explained in the following.

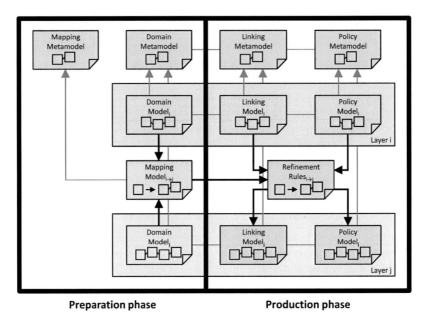

Figure 1.2: Approach

Horizontally, we represent knowledge about policies and the underlying domain in different models. Domain experts specify in the domain model which information is available in the policy-based system. The modeling of the domain is a prerequisite for the specification of policies. It is performed at design time and represents the preparation phase of the system. In a second step, policy experts specify which policies are used to control the system behavior. This can be performed either at design time or at runtime and represents the production phase of the system. For this purpose, two models are specified. The policy model describes which policies are used to manage the system. The linking model represents an interface for the policies to the domain. Policies necessarily access domain-specific information e.g. to gather information for decision making or to invoke system operations for action enforcement.

Vertically, we represent policies and the underlying domain at different abstraction layers. In figure 1.2, these layers are highlighted in blue. The number of layers is flexible and depends on the complexity of the domain. Each policy has a representation at each layer, from the highest layer with a functional focus down to the lowest layer with a technical focus. Each layer offers a particular view for a particular expert group concerning the system behavior. Initially, domain and policies are specified at the highest layer to describe system behavior already at an early stage of development, when application experts only deal with high-level aspects of the system. These initial models at the highest layer provide the starting point for the refinement, as indicated with black arrows in figure 1.2. At first, domain experts refine the domain by providing a more detailed specification of the domain model at the subsequent lower layers. In the mapping model, they describe how the domain model at a lower layer can be derived from the precedent higher layer. Then, the policies are refined from the highest layer down to the lowest layer. For this purpose, refinement rules apply the knowledge enclosed in the mapping model to the policy and linking models at the higher layer. Refinement rules represent model transformations that generate the respective policy and linking models at the subsequent lower layer in an automated way. If more than two layers are available, the refinement is performed iteratively across the intermediate layers until the policies are represented at the lowest layer. During the refinement process, the policies are extended and additional information is integrated. At the lowest layer, the policies are finally represented in a machine-executable way, and they are ready for deployment. From this technical representation, executable code in a policy language can be generated, so that the policies can be executed by a particular policy engine.

An essential benefit of the automated policy refinement is the instant reflection of changes at the high-level abstract models in their low-level technical representation and even in the generated code. This allows to manage the system behavior at a high level of abstraction and is especially useful if the respective policies are expected to undergo frequent changes.

Approach

We separate domain knowledge from system behavior, use policies to specify the desired system behavior at different levels of abstraction, and reuse existing knowledge for an automated refinement of these policies.

1. For this purpose, we introduce different levels of abstraction within the domain, and we provide a selected level to each expert group to represent domain-specific knowledge in their individual terminology.

2. We use policies at each level of abstraction to specify the reactive behavior of the system, and we let each expert group represent policies at their particular level.

3. To refine policies from a higher level to the subsequent lower level, we establish cross-level mappings between domain-specific concepts at design time, and we apply these mappings to the policies.

4. Finally, we adjust system behavior at runtime by changing the policies at the highest level and propagating these changes down to the policies at the lowest level in an automated way.

Contributions

We provide different models at different abstraction layers to specify policies and the underlying domain, and we provide model transformations to generate refined policies in an automated way at design time and at runtime.

1. A generic metamodel allows to model the domain and domain-specific concepts at different abstraction layers and offers a particular view of the relevant parts of the domain at each layer.

2. Another generic metamodel allows to model domain-specific ECA policies at the different abstraction layers and offers a particular view of the relevant policies at each layer.

3. Mapping patterns allow to specify how the domain-specific concepts at a lower layer are derived from the precedent higher layer; a generic metamodel allows to model the cross-layer mappings.

4. Refinement rules finally apply the cross-layer mappings to the policies at the highest layer and generate refined policies at the lower layers in an automated way.

1.4 Publications

Parts of this thesis have appeared in the following publications. In the list, we distinguish between full papers, short papers, and demo papers. Furthermore, we reported several inventions from which one was filed as a patent application. Publications with several authors were written in collaboration with colleagues or students who contributed to the results of this thesis.

Chapter 3 has been published in [38–40], and chapter 4 includes material from [38,41–43]. The background of the case study in section 5.1 is described in [44–49].

Full Papers

- Raphael Romeikat, Stephan Roser, Pascal Müllender, and Bernhard Bauer. Translation of QVT Relations into QVT Operational Mappings. In *International Conference on Model Transformation (ICMT)*, pages 137-151. Springer LNCS, July 2008. [41]

- Raphael Romeikat, Markus Sinsel, and Bernhard Bauer. Transformation of Graphical ECA Policies into Executable PonderTalk Code. In *3rd International Symposium on Rule Interchange and Applications (RuleML)*, pages 193-207. Springer LNCS, November 2009. [42]

- T. Bandh, G. Carle, H. Sanneck, L. C. Schmelz, R. Romeikat, and B. Bauer. Optimized Network Configuration Parameter Assignment Based on Graph Coloring. In *12th Network Operations and Management Symposium (NOMS)*, pages 40-47. IEEE Communications Society, April 2010. [44]

- Raphael Romeikat, Bernhard Bauer, Tobias Bandh, Georg Carle, Henning Sanneck, and Lars Christoph Schmelz. Policy-driven Workflows for Mobile Network Management Automation. In *6th International Wireless Communications and Mobile Computing Conference (IWCMC)*, pages 1111-1115. Association for Computing Machinery, June–July 2010. [45]

- Raphael Romeikat and Bernhard Bauer. Specification and Refinement of Domain-Specific ECA Policies. In *4th International Workshop on Domain-specific Engineering (DsE@CAiSE)*, pages 197-206. Springer LNBIP, June 2011. [38]

- Raphael Romeikat, Bernhard Bauer, and Henning Sanneck. Modeling of Domain-Specific ECA Policies. In *23rd International Conference on Software Engineering and Knowledge Engineering (SEKE)*, pages 52-58. Knowledge Systems Institute, July 2011. [39]

- Raphael Romeikat, Bernhard Bauer, and Henning Sanneck. Automated Refinement of ECA Policies for Network Management. In *17th Asia-Pacific Conference on Communications (APCC)*, pages 439-444. IEEE Communications Society, October 2011. [43]

Short Papers

- Raphael Romeikat and Bernhard Bauer. Formal Specification of Domain-Specific ECA Policy Models. In *5th International Conference on Theoretical Aspects of Software Engineering (TASE)*, pages 209-212. IEEE Computer Society, August 2011. [40]

Demo Papers

- Tobias Bandh, Henning Sanneck, and Raphael Romeikat. An Experimental System for SON Function Coordination. In *International Workshop on Self-Organizing Networks (IWSON)*, pages 1-2. IEEE Vehicular Technology Society, May 2011. [46]

- Tobias Bandh, Raphael Romeikat, Henning Sanneck, and Haitao Tang. Policy-based Coordination and Management of SON Functions. In *12th International Symposium on Integrated Network Management (IM)*, pages 823-836. IEEE Communication Society, May 2011. [47]

- Tobias Bandh, Henning Sanneck, and Raphael Romeikat. Demonstration of an Integrated SON Experimental System for Self-Optimization and SON Coordination. In *2nd International Workshop on Self-Organizing Networks (IWSON)*, to appear. IEEE Vehicular Technology Society, August 2012. [48]

Patent Applications

- Patent application, Nokia Siemens Networks Oy, June 2011. Not yet disclosed. [49]

1.5 Outline

We provided background and motivation for this thesis in chapter 1. Based on increasing complexity in the past, we identified problems with the development of IT systems and derived respective objectives. We presented an approach to achieve these objectives and summarized the respective contributions of the thesis. The chapter also provided a list of publications and summarizes the following chapters.

In chapter 2, we provide the relevant policy-based background for this thesis. The chapter defines terms and discusses concepts used in this thesis. This covers different aspects of PBM from its historical roots to various policy frameworks that are used today.

We present our approach to the specification of policies in chapter 3. To raise the level of abstraction and reduce complexity, we apply techniques from MDE. Our approach features different models at different levels of abstraction in order to facilitate the collaboration of expert groups. Furthermore, we formally define the abstract syntax of the models and illustrate the essential aspects with a running example from the network management domain.

In chapter 4, we describe the refinement of policies from an abstract into a technical representation. The refinement extends the initial policies in order to make them executable. We use model transformations to automate the refinement process as far as possible and formally define the semantics of the refinement. The running example again helps to illustrate our approach.

We demonstrate the specification and refinement of polices with two case studies in chapter 5. The first case study deals with the management of Next Generation Mobile Networks (NGMN), the second one with bonus payments for employees. Both case studies represent real-world scenarios that were provided by industry.

In chapter 6, we summarize our approach and discuss our contributions and their implications with regard to the identified challenges. Finally, the thesis concludes with an outlook on future work.

2

Policy-Based Management

Since multiple decades, Policy-Based Management (PBM) has gained attention in research and industry as a flexible approach to the management of complex systems. PBM primarily addresses the abstraction and automation of a system at runtime. It enables modifications of system behavior without considering technical details or changing source code. This is achieved with the help of policies that define which actions must or must not be performed according to the particular situation. Thus, policies represent context-sensitive rules that accomplish decisions in an automated way. Administrators can change system behavior at runtime by changing the respective policies. This allows to control and manage a system in a simple and flexible way.

A well-known application area of PBM is network management. Here, policies are widely used to automate configurations and optimize processes. We recently considered the usage of policy-based systems for the management of mobile networks in [43, 45–48, 50]. Other application areas of PBM include security management, service management, and configuration management in large-scale computer systems [51–53].

In this chapter, we present important aspects of PBM in order to ensure a proper understanding of the following chapters. For this purpose, section 2.1 reviews how policies have evolved in the past. Section 2.2 introduces basic notions about policies and their specification, refinement, and enforcement. Section 2.3 presents relevant policy languages and engines. Finally, section 2.4 provides a conclusion of the state of the art and an outlook on future work.

2.1 History

The term *policy* has its origin outside Information Technology (IT). In general, a policy is a "plan of what to do in particular situations, that has been agreed officially by a group of people, a business organization, a government, or a political party" [54]. Thus, policies represent directives that are related to a certain domain and influence certain decisions by limiting freedom of action. In this sense, Asimov's Laws of Robotics, first published in 1950, can be regarded as policies for the robotics domain [55, 56]:

- First Law: A robot must not injure a human being, or, by inaction, allow a human being to come to harm.

- Second Law: A robot must obey orders given to it by human beings, except where such orders would conflict with the First Law.

- Third Law: A robot must protect its own existence as long as such protection does not conflict with the First or Second Law.

Experts in artificial intelligence regard the Laws as a future ideal: once a system has reached a stage where it can comprehend the Laws, it is truly intelligent. However, a robot cannot understand the Laws with technology available today as they are specified in natural language with no formalism behind. For policies to be interpreted by machines, a formal and structured representation is required. In IT, policies have been applied for different purposes, and various terms and definitions have been used with a different focus [57]. In the following, we describe how PBM has evolved in the past.

2.1.1 Access Control

Increasing complexity of Information and Communication Technology (ICT) systems and the need for their autonomous management have driven research and industry to look for appropriate management frameworks. Policy-based systems were considered a promising approach. Early work on policies dating back to the 1960s focused on security aspects and primarily on access control. Access control is the process of mediating requests to resources maintained by a system and deciding for each request whether it should be granted or denied. This decision is specified by security or authorization policies, and these can be realized with different mechanisms, depending on what should and what should not be allowed [58].

- Access Control Lists (ACLs) were introduced to share objects between subjects and to protect access to these objects with an access control matrix [59–61]. Each (*subject, object*) entry in the matrix contains a set of access attributes such as *read*, *write*, or *owner* that define the access rights of the respective subject to the respective object. The matrix represents a set of security policies as it specifies how to decide on access requests of subjects to objects. The ACL approach typically partitions the matrix by columns, i.e. for each object, a list of subjects is maintained that may access the object. In contrast, the capability-based access control model partitions the matrix by rows, i.e. for each subject, a list of objects is maintained to which the subject has access [62,63]. Further considerations of security policies replaced the $(subject, object)$ duples with $(user, program, file)$ access triples [64]. The previous approaches are classified as Discretionary Access Control (DAC) as access restrictions are only based on the identity of the subject. In DAC systems, a subject has complete control over the objects it owns and may grant access to other subjects.

- In contrast, access is managed centrally in Mandatory Access Control (MAC) approaches. Security policies are specified by an administrator and cannot be changed or extended by subjects. MAC-based systems typically classify objects according to security levels (such as top secret, confidential, or public) and authorize subjects to these levels (clearance) as in the Bell-LaPadula model [65, 66]. The classification and clearance of an access request determine whether the request is granted or denied. In particular, a subject that is authorized to a particular security level must not read an object of a higher level (no read-up) and must not write to any object at a lower level (no write-down). MAC is most commonly used in systems where confidentiality and controlled access to classified information are important. It allows to implement organization-wide security policies or security policies that are guaranteed for all users.

- Role-Based Access Control (RBAC) is an approach to the management of access based on the job or function of a subject within an organization [67, 68]. It is sufficiently general to simulate both traditional methods DAC and MAC [69]. A unified model for RBAC was created later [70] and adopted as an American National Standard for Information Technology (ANSI) [71]. In RBAC, permissions are not assigned to subjects, but to roles. Each subject is then assigned to one role and thus provided with the permissions of that role. Roles can be ordered in a hierarchy to reflect the administrative hierarchy of the organization. In contrast to groups, a subject may only be assigned to one single role and cannot not be provided with any further permissions. RBAC is a flexible approach as permissions are assigned indirectly via roles. Roles have gained wide acceptance in large organizations as they can simplify access control particularly for a high number of subjects. Figure 2.1 shows that in IT-intensive industries, the majority of users' permissions is managed with RBAC today [72].

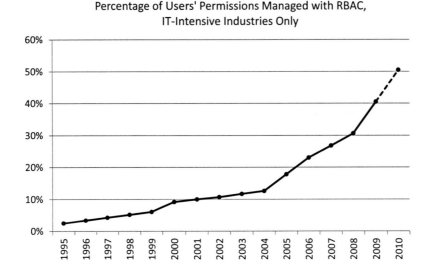

Figure 2.1: RBAC adoption [72]

2.1.2 Management Paradigm

In PBM, a more abstract perception of polices as a management paradigm
has been developed since the late 1980s. Policies were then perceived as a
technique not only to control access, but to manage distributed computing
systems in general. Here, they were considered rules to manage resources
in a scalable and efficient way. By using policies, management effort should
not increase along with an increasing complexity of a system, and an optimized usage of resources should be achieved.

- As large distributed systems may contain a very high number of resources, it is not feasible to specify policies for each resource individu-
  ally. For this reason, domains were introduced to group resources and
  to specify policies for groups of resources [73,74]. A domain represents
  a set of resources to which the same policies apply. Thus, domains
  reflect organizational boundaries and allow to partition management
  responsibility according to the organizational structure of a system. If
  necessary, domains can overlap or be nested. The usage of domains
  seemed useful to save cost, and domains also began to appear in commercial systems [75].

- Furthermore, policies were supposed to control and optimize the usage of resources in a system and thus increase its performance. An
  appropriate definition of policies is provided by [76]: at every level
  of a corporation from the corporate level across small business units

down to the technical level, policies specify the desired behavior of the underlying resources. Two levels are illustrated in figure 2.2. Policies are derived from a higher-level strategy and provide support in achieving the respective goals at the lower level. However, they only specify *what* has to be accomplished, but not *how* this is achieved.

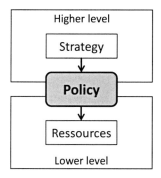

Figure 2.2: Management policies [76]

2.1.3 Policy-Based Network Management

Since the 1990s, network management has become a major application area of policies. In those days, networks were faced to an increasing number of users and Network Elements (NEs), and networks started to cross organizational boundaries via the Internet. New services were deployed, and new requirements were imposed on the underlying infrastructure. As network traffic increased, Quality of Service (QoS) became an issue. QoS does not only refer to the achieved service quality, but also deals with techniques to measure, control, and guarantee service characteristics such as bandwidth, service availability, and error rates. Due to QoS requirements and increasing network complexity, more effort was necessary for the management of these networks. Policy-Based Network Management (PBNM) was supposed to simplify management tasks with policies that monitor and configure the network and its services in an automated way [19].

- In growing networks, routing was an important issue. Routing deals with selecting an appropriate path from a source to a destination in a network in order to transmit data. Depending on QoS requirements, different data can be routed via different paths, and the obviously shortest path is often not the desired one. Although the Type of Service (ToS) field in the Internet Protocol version 4 (IPv4) header was designed for routing preferences, this field was not as widely used by Internet Protocol (IP) implementations as originally intended [77].

Clark was the first to propose a policy-based routing protocol [78]. It partitiones the network into Administrative Regions (ARs) and controls the flow of packets between ARs with policies. These policies are expressed as tuples and e.g. allow to constrain which traffic may pass without authentication from and to which IP addresses. At the same time, the Border Gateway Protocol (BGP) was published, which has been the core routing protocol of the Internet to this day [79]. BGP is primarily used for exchanging reachability information and carries information that can be used by routers for making policy-based decisions. These routing policies enforce restrictions on routing behavior and on the propagation of routing information to neighboring NEs. Furthermore, policies played an essential role in the Inter-Domain Policy Routing (IDPR) protocol [80, 81]. This protocol additionally offers source policies at the origin of a message to specify user preferences for message forwarding such as a maximum delay. Thus, IDPR was the first protocol to generate routes that satisfied user requirements and service restrictions at the same time.

- Later, the Internet Engineering Task Force (IETF) and the Distributed Management Task Force (DMTF) worked on a standardized policy framework and related protocols [82, 83]. This effort resulted in a Request for Comments (RFC) that defined terminology for PBM systems with a focus on network management, as agreed by the major vendors of such systems [84]. Here, a policy is defined as "a set of rules to administer, manage, and control access to network resources". The contained rules bind a set of actions to a set of conditions. Furthermore, a conceptual model of a policy-based system was developed, as shown in figure 2.3 [19]. In this model, the policy console represents a management tool and provides an interface to the operator for performing administrative tasks such as specification, editing, and analysis of policies. The console interacts with a policy repository, which is a physical or logical container that holds policies and related data. The repository offers means to store and retrieve policy information and can e.g. be realized as Lightweight Directory Access Protocol (LDAP) server. The Policy Decision Point (PDP) is an entity in the network that makes policy decisions based on the conditions and actions in the contained rules. Whenever a decision is requested by a subject, the PDP retrieves the respective policies from the repository, interprets them, and communicates the decision to the relevant Policy Enforcement Points (PEPs). A PEP is an entity in the network (such as a host, router, or firewall) that enforces policy decisions by executing the respective actions on the respective targets. The Common Open Policy Service (COPS) protocol was proposed for the communication between a PDP and a PEP [85], but other protocols such as the Simple Network Management Protocol (SNMP) [86] can be used as well.

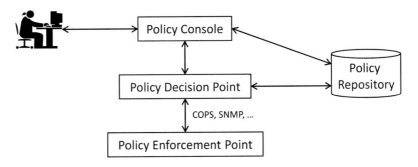

Figure 2.3: Conceptual model of a policy-based system [19]

- Since then, PBNM has been in the focus, and policies have been considered as a means to simplify and automate the network administration process [87]. The European research project SOCRATES recently focused on self-organization in Long Term Evolution (LTE) radio networks [88,89]. The idea of the project was to equip a network with the capability to automatically adapt its configuration to the environment. The resulting network is called a Self-Organizing Network (SON). In the project, 24 applications and use cases were determined that deal with self-configuration, self-optimization, and self-healing. Specific algorithmic solutions were designed and implemented as SON functions. An important consideration was that SON functions would often operate concurrently with different goals and thus conflict with each other. Conflicts are likely to arise if different SON functions target the same parameter or if a parameter is modified by one SON function and serves as input to another SON function at the same time. Such interaction may result in sub-optimal performance and user satisfaction. For this reason, a SON Coordinator framework was developed to detect and correct undesired network behavior and to ensure that the SON functions jointly work towards the same objective. Objectives are expressed by the network operator as operator policies, representing a context-sensitive strategy to resolve conflicts by prioritization, i.e. by weighting and ranking the conflicting targets [90,91]. The network operator controls the SON activities via these policies. However, the definition of operator policies remains rather unclear, and the policy enforcement remains an open issue as the project did not address techniques for the precise specification and refinement of policies.

2.2 Policies

Policies offer a promising solution to the management of complex systems, and they are used in various application areas today. In the following, we clarify the term *policy* in the context of this thesis and explain important aspects of policies. Furthermore, we describe essential steps to be undertaken for an effective development and operation of policy-based systems.

Figure 2.4 shows a typical management perspective on a policy-based system according to [92] and highlights the role of policies in this regard. Red color in the figure indicates a manual task, and green color indicates a task that is supposed to be automated. For the development of policies, developers initially specify high-level policies that describe the desired behavior of the system at a high level of abstraction. These high-level policies are then refined into a technical representation at a low level of abstraction and finally implemented in an executable language to make them applicable in the system to be managed. During their development, policies are usually stored in a repository. From there, they are deployed to the system where they should be enforced. At runtime, administrators monitor the behavior of the system in operation. They control the system and are able to change its behavior by changing the respective policies. Finally, the administrators provide feedback to the developers with respect to the developed policies and their effectiveness for the system operation. In future, an automated adaption of policies based on monitored system behavior is desirable.

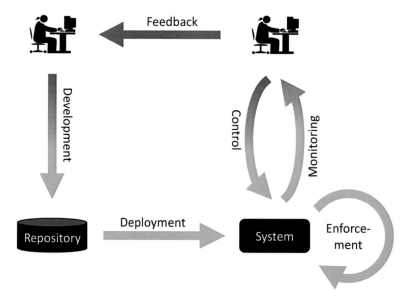

Figure 2.4: Management perspective on a policy-based system

2.2.1 Policies and Rules

Policies are used in ICT systems to separate dynamic behavior from fixed functionality. From a technical point of view, dynamic behavior is specified by declarative rules. Rules are an appropriate means to express logic that is likely to change such as logic for access control or routing purposes. The behavior of the system can then be adapted by changing the respective rules, and this should be possible at runtime without restarting or recompiling the system. According to the purpose of a policy, rules of different types can be used. Figure 2.5 shows common types that are distinguished in literature [93].

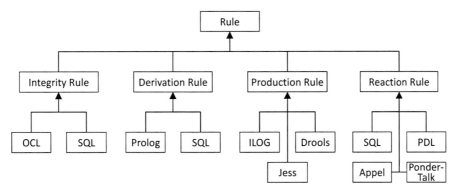

Figure 2.5: Rule types [93]

- **Integrity rules** specify constraints that must hold at anytime, e.g.: "The driver of a rental car must be at least 23 years old". Different modalities are considered to qualify a constraint. An alethic integrity rule imposes a necessity that cannot be violated at all. In contrast, a deontic integrity rule imposes an obligation that should not be violated, but theoretically could. Integrity rules are e.g. offered in the Object Constraint Language (OCL) [94] by means of invariants and in SQL [95] by means of assertions. Notably, integrity rules do not only appear in ICT systems, but also in physical, biological, and social systems, e.g.: "Every person must only be married to one other person".

- **Derivation rules** (or deduction rules) derive a set of conclusions from a set of conditions. This process is called inference. An example with one condition and conclusion is: "A gold customer is a customer with more than $ 1 million of revenue". A derivation rule without a condition is called a fact as the conclusion holds in any case. A derivation rule without a conclusion is called a denial constraint and can be used to suppress any conclusions from the respective conditions. Derivation rules have a precise theoretical foundation such as in the derivation rule system Prolog [96]. Although SQL is not a derivation rule language, it offers derivation rules by means of views.

- **Production rules** are another common type of rule that uses conditions to derive conclusions. To determine which production rules are applicable in a system, their conditions are continuously evaluated against a fact base. An example is: "Add ketchup to any order of chips". Here, an order is represented as a fact, the condition of the rule evaluates whether chips were ordered, and the addition of ketchup is the conclusion of the rule. In contrast to derivation rules, production rules usually do not have purely declarative semantics. The salience of a production rule represents its priority and adds imperative execution semantics. If multiple rules are applicable to a set of facts, the rules with highest salience are executed first. The execution sequence of rules with the same salience is usually determined by appropriate conflict resolution strategies. Production rules were widely used in the 1980s to implement expert systems and are still used in the business rules industry today. ILOG [97], Drools [98], and Jess [99] are examples of well-known production rule systems that produce conclusions from conditions.

- **Reaction rules** are considered an important rule type as they allow to describe the reactive behavior of a system in a flexible way. They specify which actions must or must not be executed in certain situations. For this purpose, a reaction rule correlates a set of events, a set of conditions, and a set of actions. The conditions are evaluated on the occurrence of the events, and they determine whether the policy is applicable or not in that particular situation. If the conditions are met, the actions are executed. For this reason, reaction rules are also called Event Condition Action (ECA) rules, ECA policies, or obligation policies. An example is: "When a share price drops by more than 5%, then sell it". Here, a share price drop could be indicated by an event containing the quantity of the drop, the condition would then check whether the quantity is more than 5%, and the action would invoke the selling of the share. Reaction rules are e.g. available in SQL by means of triggers and in dedicated languages for ECA policies such as the Policy Description Language (PDL) [100], the Adaptable and Programmable Policy Environment and Language (APPEL) [101], and PonderTalk [102]. ECA rules can be extended by adding a postcondition that accompanies the triggered action. An Event Condition Action Postcondition (ECAP) rule allows to specify the effect of a triggered action in a declarative way. The postcondition in the example above might state that the share is no more contained in the portfolio.

2.2.2 Policy Specification

In the following, we describe important aspects with respect to the specification of policies at design time. We describe the structure of the different policy types with a focus on ECA policies as these are in the focus of this thesis, and we describe how policies are represented at different levels of abstraction.

Structure

In the simplest case, a rule is already considered a policy, i.e. a policy represents one rule. Figure 2.6a shows the structure of such a flat ECA policy as a metamodel in Unified Modeling Language (UML) [103, 104]. The aggregations in the figure illustrate that a policy contains a set of events, conditions, and actions. The respective cardinalities enable policies to share events, conditions, and actions between each other, and thus to combine these in a flexible way. ECA Policies can also be defined as a more complex construct with a set of contained rules, as shown in figure 2.6b. In this notion, the contained rules are also called policy rules, and each policy rule typically has a flat structure again. Often, a contained policy structure is also called a composite policy or a policy group.

(a) Flat   (b) Contained

Figure 2.6: Metamodel of ECA policies

These flat and contained policy structures are rather generic and describe a common understanding of ECA policies. Particular policy languages or policy-based systems may use different or more complex structures that e.g. allow for nested policies, impose different cardinalities, or restrict the sharing of events, conditions, and actions between policies.

Policies that use other rule types than reaction rules will have a different structure, of course. Polices that use integrity rules will use one or more constraints, policies that use derivation rules will combine conditions and conclusions, and policies that use production rules will have a structure similar to the one of ECA policies, but not use any events.

Although different policies have different structures and use different rule types, they can often be transformed into each other while keeping their semantics. This is useful if only a particular rule type is available in a policy-based system. The following policies use different rule types, but express statements with the same meaning:

"The driver of a rental car
must be at least 23 years old."
(Integrity rule)

"If a car is rent to a driver,
the driver is at least 23 years old."
(Derivation rule)

"On an application for a rental car,
deny the rental if the driver is less than 23 years old."
(Reaction rule)

"With any application for a rental car,
check whether the driver is at least 23 years old."
(Production rule)

Abstraction

As mentioned before, policies are used to describe the behavior of a system and to abstract from technical details or aspects of implementation. In recent years, different approaches have been proposed to an abstract policy representation that is independent of any implementation language [105–111]. These approaches are summarized in section 3.6. Abstraction is important as different experts in an organization have different views of a system and use different means to describe system behavior. Top management staff e.g. express policies as business rules that describe an operative strategy to direct a system and to reach the objectives of an organization. In contrast, heads of department are concerned with the realization of these objectives, and for this purpose specify policies as guidelines and constraints on particular processes and functions. Finally, IT experts implement these policies in the particular systems of the organization.

The purpose of abstraction is to suppress details that are not of interest [112]. Each expert group specifies policies at a different level of abstraction and is not concerned with the details of the levels below. Policies at different levels of abstraction represent a policy hierarchy [113,114]. The different levels in a policy hierarchy offer different views of the policies and often represent the organizational levels in the respective organization. Figure 2.7 illustrates the characteristics of policies at different levels. High-level policies focus on business aspects and address objectives of an organization or a system as a whole. They have a rather wide scope, i.e. they potentially require a lot of functionality to be fulfilled and thus affect major parts of the underlying system. Due to their high level of abstraction, these policies lack in technical details and are therefore not enforceable. High-level policies are often expressed in natural language [115] or in appropriate formal languages such as Semantics of Business Vocabulary and Business Rules (SBVR) [116]. In contrast, low-level policies focus on technical aspects and have a rather limited scope as they only describe particular functionality. Due to their low level of abstraction, they contain technical details of the underlying system. Low-level policies are usually specified in a machine-executable language and can directly be enforced by a respective engine. If a policy hierarchy contains more than two abstraction levels, intermediate levels between the highest and lowest one allow to specify mid-level policies. According to the particular system, these policies might describe system behavior with both business and functional aspects, or they might address specific aspects or components of the system. The specific number of layers in a policy hierarchy is flexible and only depends on the particular system. Systems that are not driven by business aspects e.g. do not require any business layer. However, a higher number of layers might be reasonable along with an increasing complexity of the system. It may also be reasonable to add layers if the system evolves over time.

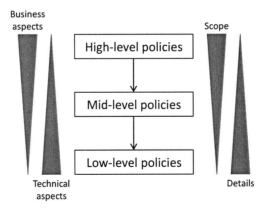

Figure 2.7: Policy hierarchy

In PBNM, different levels of abstraction are defined as the policy continuum [19, 117, 118], as shown in figure 2.8. The policy continuum represents policies from a business view down to a technical view and enables each expert group to specify and manage policies with their individual terminology. Although the number of levels is flexible and only depends on the specific system, five levels are proposed as a typical realization for network management policies.

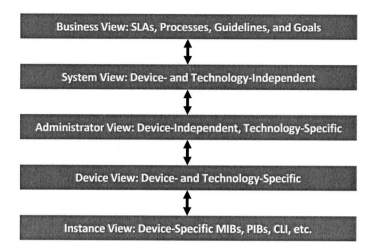

Figure 2.8: Policy continuum [19]

At the business view of the policy continuum, policies are represented with the highest level of abstraction for business users as Service Level Agreements (SLAs), processes, guidelines, and goals. An example business policy is: "Offer gold, silver, and bronze service packages to customers". This high-level policy is expressed at the system view as a set of more detailed policies, e.g.: "In the gold service package, offer up to three content services with three different service grades". At the administrator view, technology-specific commands are used to map this policy to the network architecture, e.g. resulting in a policy: "Provide a Differentiated Services (DiffServ) configuration for 6 Mbps with 50% of bandwidth for service grade 1". DiffServ is a network architecture that allows to classify network traffic to provide QoS in modern IP networks [119]. The same policy is expressed at the device view with respect to the configuration of particular network devices such as core and edge routers of the DiffServ network. Finally, the instance view uses specific commands for each device to enforce the respective configuration.

2.2.3 Policy Refinement

A high-level policy usually has little in common with the respective low-level policies. Although serving the same goal, policies at the highest level of abstraction are expressed with business terms, whereas they are specified with technical terms at the lowest level. The different views cause a semantic gap between the highest and lowest level [120]. To be executed by the underlying system, a high-level policy has to be transformed into a representation at a lower level of abstraction. This transformation is also called policy refinement. During the refinement, higher-level policies represent the basis from which lower-level policies are derived. The objective of policy refinement is to overcome the semantic gap between the highest and lowest level of abstraction, and this is a non-trivial task. Complexity arises from the fact that very detailed knowledge about the policies and the underlying system is required in order to express the same information semantically equivalent at different levels of abstraction. If more than two levels of abstraction are available in a policy hierarchy, each subsequent lower level details the information of the preceding higher level. Policies are mapped stepwise to the lowest level and made more specific as technical information is added, so that they are finally represented in a machine-executable way. A single policy may also be refined into a set of policies in order to express the same information more specifically at a technical level.

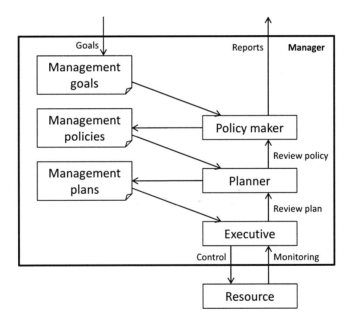

Figure 2.9: Early ideas of policy refinement [121]

Figure 2.9 shows one of the early ideas that dealt with policy refinement. The authors address the management of large scale network systems and propose one manager component per domain [121, 122]. As mentioned before, it was popular in those days to group resources into domains and domain hierarchies in order to partition management responsibility for these resources. In this context, management is considered the process by which managers monitor and control resources to meet management goals. For this purpose, a manager operates as follows. A manager receives management goals from higher-level managers. A management goal is a statement about what to achieve and can be considered a high-level policy. To achieve the received goals, the manager must refine them into management plans that can be considered low-level policies. A management plan is a sequence of actions that can be evaluated deterministically before being executed. If a management plan contributes to a management goal, the manager executes the respective actions on the managed resources. As the refinement of management goals into plans is a complex task, mid-level management policies are introduced as intermediate step. Management policies restrict the possible sequences of actions and thus simplify the refinement. Management policies are derived from the management goals by the policy maker, and management plans are derived from the management policies by the planner. For this purpose, the policy maker and the planner should have information about the capabilities of the managed resources. The authors already had the vision of an automated refinement solution, but it was not yet clear how to formally specify the different kinds of information processed within the manager.

Early ideas of policy refinement did not only consider network systems. In a more formal way, refinement mappings are used in [123] to prove that a lower-level specification correctly implements a higher-level one. The specifications require a certain system behavior, and these requirements can be considered policies on the behavior of the system. Refinement mappings create a link between system behavior according to the lower-level specification and system behavior according to the higher-level one. A state machine model is used to formalize the underlying system and its behavior, and the mappings are established between the states of the system. Precisely, a mapping links one low-level state to one high-level state. The purpose of the mappings is not to refine a high-level specification into a low-level one. Instead, an existing low-level specification is mapped to a high-level one. However, such mappings represent an important step to relate representations of behavior at different levels with each other and to formalize this relationship.

To reduce manual effort for the developer, automated solutions are desirable to policy refinement, although automation might not be practical in all scenarios [124]. If policy refinement cannot be automated, policies must be refined manually by re-writing higher-level policies in the different terminology of the lower levels. In recent years, renewed interest on this topic has been observed, and a lot of partly and fully automated approaches have been proposed. Typical approaches are summarized in section 4.3. However, these approaches are often tailored to specific applications or have major limitations. [118, 125, 126] propose approaches for a particular application domain, and [118, 127–129] can handle a fixed number of abstraction layers only. Furthermore, [128, 129] only provide partial automation and still involve a lot of manual effort.

2.2.4 Policy Enforcement

In the following, we describe important aspects with respect to the enforcement of policies at runtime. We describe how policies of the different types are triggered, and we present important detection and resolution techniques for policy conflicts with a focus on ECA policies.

Triggering

The declarative nature of policies implies a couple of benefits and e.g. allows for a high degree of parallelism in a system [130]. As policies represent declarative logic, they relieve the developer of specifying an explicit control flow. The triggering of policies is then a matter of runtime, i.e. policies are evaluated and enforced instantly and only when needed.

- Constraints of an **integrity rule** must hold at anytime in a system. Usually, it is sufficient to trigger the checking of these rules when the underlying data source is changed. Invariants in OCL must be checked when the respective model is changed, and assertions in SQL must be checked when the respective table or database is changed.

- **Derivation rules** are used to derive implicit data from explicit data. In Prolog, they are used to determine whether a query can be satisfied with a given fact base. A query is a logical statement to be proved right or wrong according to the fact base and the derivation rules. When a query is posted by a user, the Prolog engine applies the rules in order to answer the query. Here, a typical strategy is starting with the query and working backwards via subqueries to find the supporting facts (backward chaining). A view in SQL can be regarded as a derivation rule to gather, compose, and present data from a database. To keep an SQL view always updated, the query should be triggered when the view is accessed by a user or if the underlying data in the database is changed.

- In contrast, the purpose of **reaction rules** is to enforce actions according to the current situation. As these situations are explicitly specified by the occurrence of events, reaction rules are triggered and only triggered when the respective events occur. This applies to triggers in SQL and APPEL and to events in PDL and PonderTalk. Reaction rule engines typically do not actively look for events, but use a notification mechanism and remain passive as long as no events occur.

- **Production rules** represent an alternative way of handling the control flow of a system with conditional expressions, in contrast to a sequential way of programming with imperative constructs such as statements or loops. In expert systems such as ILOG, Drools, and Jess, production rules are used to generate explicit knowledge from implicit knowledge. Here, a typical strategy is starting with the facts and applying the rules to find all possible conclusions until no more rules can be applied (forward chaining). When new facts are added, the rules are applied in the same way again.

Conflict Detection and Resolution

In policy-based systems, conflicts within the set of policies are likely to occur. Two or more ECA policies are conflicting if they trigger actions at the same time that are contradictory, i.e. that cannot be executed together. The likeliness of conflicts increases along with the number of policies and the complexity of the policy conditions. There are different types of conflicts. A static conflict can be detected at design time from the specification of the policies. The following policies provide an example of a static conflict:

"On an application for a rental car,
deny the rental if the driver is less than 23 years old."
(Reaction rule 1)

"On an application for a rental car,
approve the rental if the driver is at least 18 years old."
(Reaction rule 2)

The conflict can be predicted to occur in particular situations, i.e. when a driver whose age is between 18 and 22 years applies for a rental car. The purpose of conflict detection at design time is to indicate the conflict to the developer, so that the conflict can be resolved by changing the respective policies. However, not all conflicts can be detected at design time. Reasons for this include policy conditions that depend on external input, policy actions that are triggered indirectly in an indeterministic sequence, or missing information about the underlying system. Conflicts that can only be detected at runtime are called dynamic conflicts. An example of a dynamic conflict is provided by the following policies:

"No Internet access is allowed for employees at working time."
(Integrity rule 1)

"Internet access is allowed for consultants at anytime."
(Integrity rule 2)

In this example, a predictive conflict detection is not possible if no information is available at design time whether consultants are employees or self-employed. For any static conflicts that were not resolved at design time and for all dynamic conflicts, conflict detection and resolution are important at runtime. The purpose of conflict detection at runtime is to prevent the system from executing conflicting actions that might result in an inconsistent state or even in a breaking of the system. As soon as a conflict occurs at runtime, a conflict resolution mechanism should deal with them, i.e. find an appropriate solution itself or request a manual decision. In the following, we present different conflict detection and resolution techniques for ECA policies.

- Conflicts between policies are detected by an analysis of their conditions in [131]. Two policies are potentially in conflict if their conditions can simultaneously be true. In this case, both policies may become applicable at the same time. The authors do not analyze whether the respective actions do conflict at all, but focus their analysis on the conditions, which are represented as boolean expressions. In general, it is an NP-complete problem to determine whether a conjunction of two boolean expressions is satisfiable. For this reason, the authors focus on particular subclasses of boolean expressions for which the satisfiability problem is tractable, and they provide respective algorithms for the detection of potential conflicts. However, no automated solution to conflict resolution is offered. Instead, the authors assume that potential conflicts are resolved manually by administrators. Proposed resolution techniques include the deactivation of policies and the assignment of priorities to policies.

- The traditional ECA model is extended to an Event Condition Precondition Action Postcondition (ECPAP) model for purposes of conflict detection and resolution in [132, 133]. Actions are annotated with logical preconditions and postconditions that describe the state of the system before and after the respective action is being executed. The preconditions can be regarded as an additional condition that must hold for the action to be executed. The usage of postconditions allows to detect conflicts between policy rules due to the effects of their actions. A conflict detection algorithm determines the set of rules whose events and conditions match, i.e. whose actions are triggered concurrently. Then, it checks the respective postconditions and concludes a conflict if their conjunction is not satisfied. Furthermore, the authors consider that cycles may occur when actions trigger other policy rules and provide a cycle detection algorithm. Conflicts between policy rules are resolved by resolution rules that prioritize rules by specifying preferred postconditions. In case of a conflict, the conflict resolution algorithm filters the subset of rules with prioritized postconditions. If multiple conflicting rules still remain, the resolution is ambiguous, and the developer is notified. A set of resolution rules is called a resolution policy. As a resolution policy represents a statement about other policies, it is also called a meta policy.

- Conflicts are resolved automatically with detailed precedence relationships in [134]. Several principles are used for establishing the precedence. The choice which principles to apply in which sequence is left to the system developer. The first principle is precedence based on modality, i.e. negative policies prohibiting the execution of an action always take precedence over positive ones allowing the execution of an action, or vice versa. Precedence for negative polices is commonly used with security policies to ensure that a forbidden action will never be permitted. Secondly, policies can be assigned explicit priority values to define a precedence ordering. However, this principle might result in inconsistent priorities or in arbitrary priorities that do not relate to the importance of the policies. The most sophisticated principle is the distance between a policy and the target object to indicate the relevance of the policy. Here, the idea is to give precedence to more specific policies, i.e. to policies that are closer to the target object in the inheritance hierarchy of domains or objects to be managed. According to this principle, an access control policy for a particular user group e.g. overrides an access control policy for generic users.

- A conflict resolution mechanism that filters conflicting actions is presented in [135]. In this mechanism, monitors are assigned to the policies and observe the triggered actions. If several conflicting actions are triggered concurrently, the monitors of the respective policies cancel some of the conflicting actions to obtain a set of remaining actions without any conflicts. The core of the resolution mechanism is the selection of actions to be canceled. Actions are canceled according to two principles. Firstly, as few actions as possible are to be canceled altogether. Secondly, the remaining actions must be caused by a subset of the events that caused the conflict. If an action is canceled that was caused by a certain event, any other action caused by the same event is canceled as well, unless it is also caused by a different event. Having resolved the conflict, the system behaves as if the events that caused the conflict would not have occurred at all.

2.3 Policy Languages and Engines

There is a wide range of languages available to implement policies and engines to enforce them. A policy language provides a concrete syntax, i.e. a notation for policies. Often, the concrete syntax is a textual one, as with general-purpose programming languages. Policy languages with a graphical concrete syntax are also called modeling languages. To enforce policies that are implemented in a particular policy language, an engine is required that can interpret this language. If no respective engine is available, the policies must be transformed into another language for which an engine is available. We now present an overview of policy languages that are relevant in the context of this thesis, i.e. that can be used to implement ECA policies. We also describe how the respective execution engines enforce policies.

2.3.1 PDL

The Policy Description Language (PDL) is a simple but expressive language that was developed at Bell Laboratories to be used in software switches for telephone communication networks [100, 136]. It defines a policy as a function that maps a series of events to a set of actions in order to manage the system behavior.

In PDL, a policy is a collection of different propositions. The first type of proposition is the policy rule proposition. This type corresponds to ECA rules and represents an expression to describe that an action is executed if an event occurs under a condition. The concrete syntax is:

```
event
    causes action
    if condition.
```

For this purpose, a set of events should be available that the system is able to monitor. An event can contain attributes, and each attribute of an event instance is assigned a value. An action represents functionality of the system that can be invoked by policies. A condition is a boolean expression and can refer to the attributes of the event and to available functions. A function can be invoked by the condition to evaluate the state of the system. Functions and actions can take arguments to pass information. Instead of invoking an action, a policy can also trigger another event. This type of proposition is called a policy-defined event proposition. The concrete syntax is:

```
event1
    triggers event2
    if condition.
```

The power of PDL comes from the expressiveness of the event part. Here, different types of events are distinguished. The smallest unit is a primitive event, i.e. an event is an event that simply contains attributes. A basic event represents the occurrence of one or more primitive events at the same time. A group of basic events represents all instances of a basic event at the same time. Furthermore, events can be combined to complex events that describe a particular sequence and number of the contained events.

An example policy is provided for overload control in a software switch [100]. Overload is characterized as an excessive ratio of timeouts compared to the calls made. If the timeout ratio exceeds a value t, an overload is detected. In this case, some call requests must be rejected until the timeout ratio falls below a reasonable value t' again. The respective policy is specified with four propositions as follows:

```
1   normal_mode, ^(call_made|time_out)
2     triggers restricted_mode
3     if Count(time_out) > t * Count(call_made).
4   restricted_mode
5     causes restrict_calls.
6   restricted_mode, ^(call_made|time_out)
7     triggers normal_mode
8     if Count(time_out) < t' * Count(call_made).
9   normal_mode
10    causes accept_all_calls.
```

The switch is in normal mode initially, and this is indicated by the event normal_mode. The first proposition (lines 1-3) is triggered by a complex event (line 1) describing that an instance of the normal_mode event is followed by a sequence of instances of the events call_made and time_out. The event restricted_mode (line 2) is triggered when the condition (line 3) is true for the first time, i.e. when the ratio of timeouts compared to the calls made exceeds the value t. The second proposition (lines 4-5) is triggered by the primitive event restricted_mode (line 4) and invokes the action restrict_calls (line 5). Similarly, the third proposition (lines 6-8) is triggered by a complex event (line 6) describing that an instance of the event restricted_mode is followed by a sequence of instances of the events call_made and time_out. The event restricted_mode (line 7) is triggered when the condition (line 8) is true for the first time, i.e. when the ratio of timeouts compared to the calls made falls below the value t'. The fourth proposition (lines 9-10) is triggered by the primitive event normal_mode (line 9) and invokes the action accept_all_calls (line 10).

The authors of PDL do not provide an engine to enforce policies. However, the formal semantics of PDL represents the basis for implementing a respective engine. According to the formal semantics, policies are interpreted over event histories. An event history is a sequence of zero or more epochs, and an epoch is a set of primitive event instances that occur simultaneously. The granularity of time for the epochs depends on the particular application. The formal semantics of PDL is defined in Horn logic [137] and represents a function that takes an epoch as input and determines the set of actions invoked by the policies as output. A conflict resolution mechanisms is provided that filters conflicting actions [135], as presented in section 2.2.4.

2.3.2 APPEL

The Adaptable and Programmable Policy Environment and Language (APPEL) was developed in the ACCENT project, which dealt with call control for telecommunications services [101]. Here, policies are used to specify how a system should automatically behave in certain situations. Despite the focus on telephony, the language can be applied to various domains such as Service-Oriented Architectures (SOAs) and sensor networks [138, 139]. A core language defines generic constructs such as triggers, conditions, and actions and can be extended with domain-specific constructs.

Different types of policies are considered in APPEL. Regular policies correspond to ECA rules and invoke actions in response to triggers. To simplify the parsing and interchange of policies, the concrete syntax is based on Extensible Markup Language (XML) and supported by ontologies in Web Ontology Language (OWL). The typical structure of a regular policy is:

```
<policy applies_to=... enabled=...>
  <preference>...</preference>
  <policy_rule>
    <trigger>...</trigger>
    <condition>...</condition>
    <action>...</action>
  </policy_rule>
</policy>
```

A policy is the main element in the language. Attributes within this element specify further details such as the entity to which the policy applies or whether the policy is enabled or not. The modality of the policy can be specified in form of a preference from strongly positive (action must be invoked) to strongly negative (action must not be invoked). The modality is primarily used for conflict resolution. A policy rule contains a trigger, a condition, and an action. If its trigger occurs and its condition is satisfied, the policy rule invokes its action. Policy rules can be combined or nested in order to specify more complex ways of application. Triggers represent external events such as an incoming call. A trigger has a name and can contain optional arguments to provide information to the policy rule. Multiple triggers can be combined to require either one or all triggers to occur. If no trigger is specified, the policy rule acts as a goal. Conditions are boolean expressions that depend on information derived from the trigger. A policy rule can combine multiple conditions to more complex conditions, and it can omit a condition, i.e. not depend on any information provided by a trigger. An action is obligatory, has a name, and can contain optional arguments to pass information. Multiple actions can be combined and invoked in a particular sequence.

The following policy provides an example for call control [101]. It specifies that emergency calls must not be forwarded at all. For this purpose, a regular policy is specified (lines 1-12). The policy is enabled (line 1) and applies to any user in the domain @example.com (line 1). The policy contains one policy rule (lines 3-11). Incoming calls are represented by the trigger connect, which triggers the policy rule (line 4). In the policy condition (lines 5-9), a check is performed whether the parameter call_type of the trigger equals to the value emergency. The policy refers to the action forward_to (line 10). Due to the preference must_not (line 2) of the policy, the action must not be invoked if the condition is met. This should apply without regard to the desired forwarding address, and for this reason the forwarding address is omitted in the respective argument arg1 of the action (line 10).

```
1   <policy applies_to="@example.com" enabled="true">
2     <preference>must_not</preference>
3     <policy_rule>
4       <trigger>connect</trigger>
5       <condition>
6         <parameter>call_type</parameter>
7         <operator>eq</operator>
8         <value>emergency</value>
9       </condition>
10      <action arg1="">forward_to(arg1)</action>
11    </policy_rule>
12  </policy>
```

Originally, APPEL only had informal semantics, but formal semantics were defined later at least for a slightly reduced subset of the language [140]. The formal semantics is based on temporal logic and allows to detect conflicts between policy actions. Resolution policies are used to specify under which conditions and in which way a policy server should react to conflicts and determine a compatible set of actions [141]. Furthermore, a policy server was implemented in Java to store, deploy, and enforce APPEL policies and thus provides an appropriate engine [142]. Tool support for the specification of policies and detection of static conflicts is available as well [143].

2.3.3 PonderTalk

Ponder2 is a generic system that realizes a Self-Managed Cell (SMC) [102]. An SMC represents an autonomous administrative domain and contains components that are capable of self-management. In Ponder2, these components are called managed objects and are organized in a domain hierarchy. Managed objects interact with each other through events, which represent asynchronous messages. Individual managed objects can be implemented and integrated in the SMC as Java objects and can serve as adapters for real-world objects. Policies represent a predefined type of managed objects. They are triggered by events and perform management activities on the managed objects of the SMC. Thus, policies provide a local closed-loop adaptation for the SMC. As policies are managed objects themselves, they can act as subjects and send events to manage other managed objects, or they can act as objects and be managed by other managed objects. The forerunner Ponder was designed for network and systems management [144]. In contrast, Ponder2 is an extensible framework that addresses ubiquitous systems [145], but can be applied to arbitrary domains. Its applicability was e.g. shown for healthcare systems and mobile ad-hoc networks [146,147].

PonderTalk is the language to specify policies in Ponder2. Its concrete syntax is based on Smalltalk [148] and enables message passing between managed objects. Internally, PonderTalk is compiled to XML code that is interpreted at runtime. ECA polices are called obligation policies in PonderTalk. Furthermore, authorization policies can be used to permit or deny the sending and receiving of messages. The distinction between obligation and authorization policies was already suggested by the authors in early thoughts about policy-based systems [149]. The typical structure of an obligation policy is:

```
policy := root/factory/ecapolicy create.
policy event: ...;
       condition: [ :arg1 | ... ];
       action: [ :arg1 | ... ];
       active: true.
```

If a policy is active, it receives events of the specified type whenever these are sent by a managed object. Events are indicated with their name and can contain arguments, and these arguments can be used within the condition and action blocks. Having received an event, the condition is evaluated. The condition block contains a boolean expression in its right part and refers to the used arguments in its left part. If the condition is met, the action is executed. The action block contains an action or a sequence of statements in its right part and again refers to the used arguments in its left part.

The following example shows an obligation policy for managing heart rates in a body-sensor network. This policy raises an alarm if a heart rate is measured that exceeds a critical value. For this purpose, an obligation policy is created at first (line 1). The event, condition, and action of the policy are specified in the following (lines 2-5). The policy is triggered by events of the type root/event/heartRate (line 2). The event arguments oldValue and newValue indicate how the heart rate has changed since the previous measurement. In the condition, these arguments are used (line 3) to check whether the measured heart rate goes beyond a critical value of 130 (line 4). In this case, the policy invokes the operation setAlarm and shows the alarm in its action (line 5). Finally, the policy is activated (line 6).

```
1  policy := root/factory/ecapolicy create.
2  policy event: root/event/heartRate;
3        condition: [ :oldValue :newValue |
4                    (newValue > 130) & (oldValue <= 130) ];
5        action: [ root/alarm setAlarm: true; show. ];
6        active: true.
```

An algebra is provided as preliminary work on the formal semantics of Ponder2 policies [150]. This algebra can be used to detect modality conflicts within a set of policies. A modality conflict occurs when a policy action is to be executed but denied by an authorization policy, or when two authorization policies are contradictory. Furthermore, an analysis framework is provided to reason about properties of the policy-based system [151]. Modality conflicts can be detected with this framework as well. The framework is implemented in Event Calculus (EC) [152] and provides a mapping for Ponder2 policies. It also includes a prototype analysis tool. In its execution engine, Ponder2 resolves modality conflicts between authorization policies based on domain nesting precedence [153]. The resolution mechanism primarily gives precedence to the policy that applies to more specific instances of the subject and the target of the requested policy action. Conflicts between obligation policies are not addressed in Ponder2 as the effect of executed actions is not considered. Finally, Ponder2 has a built-in shell that allows to directly interact with managed objects in the SMC. PonderTalk statements at the shell can e.g. be used to activate or deactivate policies at runtime.

2.4 Conclusion and Outlook

PBM has a long tradition in network management and other areas. It provides appropriate means to manage complex systems in a flexible way and with a high degree of abstraction and automation. Policies enable automation as they accomplish decisions without any human intervention. Nevertheless, the policy-based system is still under human control as system behavior can be changed by changing the respective policies. These changes can even be applied at runtime without the need to stop and restart the system. Unexpected issues can be handled with conflict detection and resolution mechanisms in an automated way. A variety of policy languages and engines are available to specify and enforce policies. Policies can also be modeled independently of any implementation language as abstract models. Abstraction is important for the developer as technical details are suppressed that are not of interest. Moreover, the representation of policies at higher abstraction levels provides different views of system behavior for different expert groups in an organization.

However, policies at high levels of abstraction must be refined into lower levels in order to make them applicable in the respective system. Refinement is a complex issue and often requires a lot of human effort to manually re-write higher-level policies at lower levels of abstraction. Although ideas of refinement automation have existed for a long time, automated solutions often have limitations or are tailored to specific applications. For this reason, techniques are desirable that automate the policy refinement process, and these techniques should be applicable for arbitrary policy hierarchies and continua with respect to the number of abstraction levels and the application domain. Finally, executable code should be generated from refined policies in an automated way, and this should be possible for different policy languages. Ideally, system behavior can be changed at runtime by changing the high-level models and automatically propagating these changes down to the respective code in the implementation of the underlying system.

3

Model-Based Policy Specification

In software and systems engineering, a paradigm shift can be observed towards generic model-driven approaches. With the advent of the Model-Driven Engineering (MDE) approach [14], models are not only used for discussion and documentation purposes, but they are used as primary artifacts from which implementations are generated. This moves the focus from a code-centric to a model-centric point of view and has major consequences on the way information systems are built and maintained. MDE is a global trend and a promising approach that has already proven benefits regarding cost reduction and quality improvement [154].

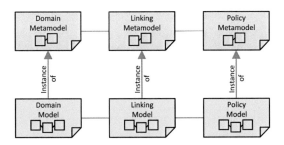

Figure 3.1: Metamodels and models

Models raise the level of abstraction and reduce complexity by separating different concerns of a system from each other. For this purpose, we present a model-based policy specification approach in this chapter that uses different models for different aspects, as illustrated in figure 3.1. In the domain model, domain experts specify which concepts are available in a system. Then, in the policy model, policy experts specify which policies are used to manage the system. Finally, in the linking model, policy experts specify where domain-specific concepts are used within the policies. The separation of domain and policy aspects is essential in order to have one generic modeling approach for different application domains and nevertheless use domain-specific concepts within the policies [155]. The domain model is specified once at design time by domain experts. Afterwards, the policy and linking models are specified at design time by policy experts. In contrast to the domain model, these two models can also be changed at runtime whenever the system behavior should be changed. For this reason, the separation of domain and policy aspects is essential. The policy and linking models could also be merged into one model only, but they are considered separate models for the sake of clarity. Each model is an instance of a respective metamodel. Common entities in the metamodels define how to put model instances together [156]. Each metamodel defines the abstract syntax of the respective model, i.e. its structure. The abstract syntax is independent of any concrete syntax, i.e. notation. As with the models, the three metamodels are actually parts of one large metamodel. Some entities appear in all three metamodels and represent the connection points between the single metamodels.

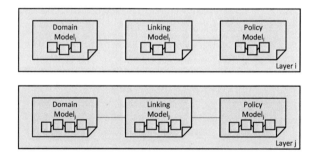

Figure 3.2: Abstraction layers and views

Policy-based management is a layered approach that represents policies at different levels of abstraction. For simple systems, it might be sufficient to have two levels only, one with a functional view and another one with a technical view. For larger systems or systems in a complex domain, it might be reasonable to introduce additional levels between the functional and the technical level in order to represent the system behavior at intermediate abstraction levels. The highest layer represents the domain and the domain-specific policies from a functional point of view, whereas the lowest layer provides a representation from a technical point of view, so that the policies can be transformed into executable code.

Similarly, the authors of [157] refer to the Model Driven Architecture (MDA) [15] and consider three different abstraction levels for rule-based approaches. The Computation-Independent Model (CIM) typically uses a natural, semi-formal, or visual language to express policies. In the Platform-Independent Model (PIM), policies are specified as statements in some formalism or computational paradigm that can be directly mapped to a particular platform. The Platform-Specific Model (PSM) implements policies in an executable language. Five levels of abstraction are proposed in the policy continuum [19,117,118], allowing to specify policies at each level in a domain-specific terminology and to refine them from a functional high level down to a technical low level.

In contrast, our approach proposes a flexible number of abstraction layers. We enable multiple abstraction layers by providing views of the models. At any layer, we define a view that filters the visible objects of the layer. These views are available for the domain, policy, and linking models. Figure 3.2 exemplarily shows two layers i and j with layer i providing a higher level and layer j providing a lower level of abstraction. The views at layer i are indicated with $DomainModel_i$, $PolicyModel_i$, and $LinkingModel_i$, and the views at layer j with $DomainModel_j$, $PolicyModel_j$, and $LinkingModel_j$. It is the task of system architects to find an adequate number of layers for a particular system. A higher number of layers partitions the semantic gap between the highest and the lowest layer into a higher number of smaller semantic gaps. On the one hand, this increases the effort for the policy refinement, but makes it more straightforward on the other hand. This consideration motivates the flexible number of abstraction layers in our approach, which represents each level of abstraction with one particular layer.

In the following sections, we describe how we model domain-specific policies at different levels of abstraction. For this purpose, section 3.1 provides a scenario from the network management domain that we use as running example throughout this thesis. Section 3.2 presents a relational algebra to formally define models. Then, section 3.3 describes the modeling of a domain, section 3.4 the modeling of policies, and section 3.5 the linking of the policies to the domain. Finally, section 3.6 presents related work, and section 3.7 provides a conclusion and an outlook.

3.1 Example Scenario

The signal quality of wireless connections in a communication system is subject to frequent variations. There are various reasons for variations with impact on the transmission, such as position changes of cell phones or changing weather conditions, as illustrated in figure 3.3. One possibility to react to varying signal quality is adjusting the Radio Transmission Power (TXP) as this setting proportionally influences the signal quality between radio antennas and cell phones.

Figure 3.3: Example scenario

From a management perspective, this scenario raises two important objectives. On the one hand, the transmission power of the antenna should be rather high in order to ensure good signal quality and avoid connection losses. On the other hand, the transmission power should be rather low in order to avoid unnecessary power consumption. As the transmission power also influences the coverage area of the respective cells, a setting too high or too low can result in an undesired state of the network [47]. Therefore, a good tradeoff between transmission power and signal quality is desired.

In order to manage the behavior of the communication system, we introduce a policy-based approach. Two Event Condition Action (ECA) policies *lowQuality* and *highQuality* are responsible for adjusting the TXP of the antenna. From a conceptual point of view, the TXP of an antenna can be increased and decreased. Changes in the signal quality are indicated by an event that contains the International Mobile Equipment Identity (IMEI) as unique identifier of the respective cell phone and the old and new value of the signal quality. The two policies are triggered whenever this event occurs, and in their conditions, they check the values of the signal quality enclosed in the event. If the signal quality falls below a critical value, the policy *lowQuality* increases the TXP of the antenna. The other way round, the policy *highQuality* decreases the TXP if the signal quality goes beyond another critical value.

The operator can then change the behavior of the communication system at runtime via the two policies. The accepted range of signal quality between the two critical values is specified in the policy conditions. Changing the critical values is possible at any time and immediately has the desired effect on the transmission power and the signal quality.

3.2 Relational Algebra

In the following sections, we formally define the domain, policy, and linking metamodels, i.e. the abstract syntax of the domain, policy, and linking models. This formalization serves different purposes. Firstly, it provides an exact definition of the models that is independent of any concrete syntax. Secondly, it allows to validate the static semantics of model instances, i.e. to check whether these are valid instances of the respective metamodels. Last but not least, the formalization represents the basis for defining the formal semantics of the refinement process later.

Following the relational model proposed by Codd [158,159], we perform the formalization with a relational algebra. This algebra represents the entites in the domain, policy, and linking models as relations and the instances of these entities as tuples in these relations, as shown in definition 1. Another benefit of a relational algebra is the possibility to persistently store the models. A persistence layer that uses a relational database can directly create a database schema from the relations and map the models to the tables of the database.

Definition 1 (*Relational algebra*)

Let Σ be a non-empty finite set called alphabet. Elements of Σ are called symbols. A word over Σ is any finite sequence of symbols from Σ. The length of a word is the number of symbols in the word. The set of all words over Σ of length n is denoted Σ^n. The set of all words over Σ of any length is the Kleene closure of Σ and denoted $\Sigma^* := \bigcup_{n \in \mathbb{N}} \Sigma^n$.

The sets

$$Val_T := \Sigma^* \tag{3.1}$$

$$Val_N := \mathbb{R} \tag{3.2}$$

$$Val_B := \{true, false\} \tag{3.3}$$

$$Val := Val_T \uplus Val_N \uplus Val_B \uplus \{undef\} \tag{3.4}$$

contain all possible values that can be used as textual, numeric, and boolean data. A value can also be undefined.

We define a model

$$M := (E_1, ..., E_k) \tag{3.5}$$

as a k-ary tuple of k entities $E_1, ..., E_k$.

We define an entity

$$E(c_1, ..., c_k) \subseteq \mathcal{P}(Val_1) \times ... \times \mathcal{P}(Val_k)$$
$$\text{where } \forall i \in \{1, ..., k\}. Val_i \subseteq Val \tag{3.6}$$

as a k-ary relation with a name E and k named columns $c_1, ..., c_k$. An entity is a finite set of k-tuples over sets of all possible values Val. A tuple represents an instance of the entity E, i.e. an object in the model M.

We define a view V of a model M

$$M_V := (E_{1_V}, ..., E_{k_V}) \subseteq M = (E_1, ..., E_k) \tag{3.7}$$

as a subset of M, i.e. the view filters some desired objects of the model. The filtering is performed by filtering the desired objects of the k entities of the model $E_1, ..., E_k$. Thus, the view is a k-ary tuple of views $E_{1_V}, ..., E_{k_V}$.

We define a view V of an entity E

$$E_V(c_1, ..., c_k) \subseteq E(c_1, ..., c_k) \tag{3.8}$$

as a subset of E, i.e. the view filters some desired objects of the entity. Thus, the view of an entity with k named columns $c_1, ..., c_k$ is again a k-ary relation with the same named columns.

The set

$$Id := Id_{La} \cup Id_{Co} \cup Id_{Pr} \cup Id_{Op} \cup Id_{Pa} \cup Id_{Re}$$
$$\cup\ Id_{Po} \cup Id_{Ev} \cup Id_{Cd} \cup Id_{Ac} \cup Id_{OI} \subseteq Val \tag{3.9}$$
$$\text{where } Id_{La} \subseteq \mathbb{N}$$

contains all identifiers that can be used to uniquely identify objects. The disjoint subsets contain all identifiers that can be used for layers, concepts, properties, operations, parameters, relationships, policies, events, conditions, actions, and operation invokings.

The set

$$Lit \subseteq Val \tag{3.10}$$
$$Lit \cap Id = \varnothing \tag{3.11}$$

contains all values that can be used as literals. Literals must not be used as identifiers and vice versa.

3.3 Domain Modeling

Different expert groups are involved in the development and operation of a system such as business managers or system administrators. Depending on their focus and their background, members of an expert group have a particular view of the system, and they use individual terminology to describe their knowledge. The domain represents a common understanding of these expert groups and gives a specific meaning to the relevant terminology. It covers knowledge about the underlying system at the different abstraction layers.

We cover any relevant information about the domain with the domain model, as illustrated in figure 3.4. The domain model specifies which domain-specific concepts are available at which layers. Its purpose is to specify any domain knowledge that should be used within policies, but to specify this knowledge independently of any policies. Policies can then use the domain-specific concepts in their event, condition, and action parts to control a system in that domain. Domain experts specify the domain model once at design time. The domain model is an instance of the domain meta-model and offers a particular view at any layer. Views filter the part of the domain model that is visible at the respective layer. Two layers i and j are shown exemplarily in figure 3.4.

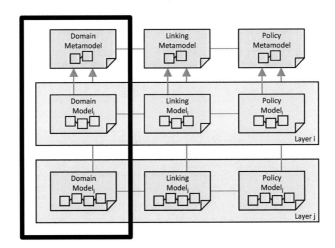

Figure 3.4: Domain modeling

In the following sections, we describe how we use the domain metamodel and the domain model to specify domain knowledge. A relational algebra formally defines the domain metamodel, i.e. the abstract syntax of the domain model.

3.3.1 Domain Metamodel

The domain metamodel allows to specify domain models in a way that is more expressive than just a domain-specific vocabulary and that is close to the structure of an ontology. For this purpose, the metamodel represents domain-specific knowledge, as shown in figure 3.5. It represents the abstract syntax of the domain, i.e. it defines the structure of the domain model.

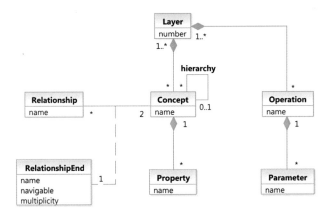

Figure 3.5: Domain metamodel

The domain metamodel contains the following entities. *Concepts* and *operations* in the domain metamodel have a name and are assigned to *layers*. These layers specify the levels of abstraction at which concepts and operations can be modeled. Each layer contains an arbitrary number of concepts and operations, and each concept and operation is assigned to one layer at least. A concept can have named *properties* assigned, and each property belongs to one concept. An operation can have named *parameters*, and each parameter belongs to one operation. Any two concepts can be assigned to each other with named *relationships*. The connection points between a relationship and a concept are called *relationship ends* and have a name, a navigability, and a multiplicity.

The domain metamodel enables tree-like hierarchies of concepts. A concept can be assigned other concepts as *subconcepts* and is called their *superconcept* in this case. A subconcept inherits the properties and relationships from its superconcept, so that any property and relationship of a superconcept is also available at its subconcepts. However, a subconcept can extend its superconcept with additional properties and relationships. A subconcept does not inherit the layers from its superconcept, and this allows to specify additional subconcepts at lower layers that are not available at higher layers. If subconcepts also inherited the layers from their superconcept, they would be available at the higher layer as well, and this is not intended.

A layer is unambiguously identified by its number and this number is used to denote the layer. Therefore, the number of a layer is unique. Layers are ordered in a top-down way, starting with *1*. $la_1 < la_2$ means that layer la_1 contains less technical details and thus represents a higher level of abstraction than layer la_2, and vice versa. Concepts, properties, operations, parameters, and relationships are unambiguously identified by their name and their name is used to denote them. Therefore, their names are unique.

3.3.2 Domain Model

The domain model specifies the available domain-specific concepts, properties, operations, parameters, and relationships at various abstraction layers. As the domain model is an instance of the domain metamodel, various restrictions must hold on its structure.

We formally define the abstract syntax of the domain model with a relational algebra in the following. Figure 3.6 shows an overview of the relations as an Entity Relationship (ER) diagram. Each relation is shown with its name and the names of its columns. Each tuple within a relation is uniquely identified by one column or a combination of multiple columns. These columns are called the Primary Key (PK) of the relation. A tuple can also reference other tuples in other relations. Such a reference is called a Foreign Key (FK) and contains the PK of the referenced tuple in order to ensure a unique reference. Foreign key references between relations are indicated with arrows in the ER diagram.

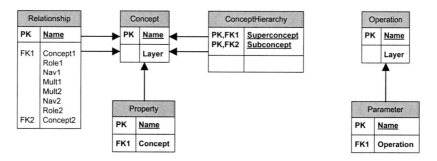

Figure 3.6: ER diagram of the domain model

Definition 2 (*Domain model*)

Let *Concept* be a relation of concepts, *ConceptHierarchy* a relation of concept hierarchy, *Property* a relation of properties, *Operation* a relation of operations, *Parameter* a relation of parameters, *Relationship* a relation of relationships.

We define the domain model

$$DM := (Concept, ConceptHierarchy, Property,$$
$$Operation, Parameter, Relationship) \tag{3.12}$$

as a sextuple of relations that specifies which concepts, concept hierarchy, properties, operations, parameters, and relationships are modeled. A domain model is valid if the model is well-formed according to (3.14), (3.19), (3.26), (3.35), (3.40), and (3.49), if the semantic properties hold for it according to (3.20)–(3.22), (3.27)–(3.28), and (3.41)–(3.42), and if (3.50)–(3.52) hold for the enclosed relations.

Let furthermore $Concept_{la}$ be a view of $Concept$, $ConceptHierarchy_{la}$ a view of $ConceptHierarchy$, $Property_{la}$ a view of $Property$, $Operation_{la}$ a view of $Operation$, $Parameter_{la}$ a view of $Parameter$, $Relationship_{la}$ a view of $Relationship$ at layer la.

We define the view of the domain model DM at layer la

$$DM_{la} := (Concept_{la}, ConceptHierarchy_{la}, Property_{la},$$
$$Operation_{la}, Parameter_{la}, Relationship_{la}) \tag{3.13}$$

as a sextuple of views that specifies which concepts, concept hierarchy, properties, operations, parameters, and relationships are modeled at a particular layer. A view of the domain model is valid if the view is well-formed according to (3.14), (3.19), (3.26), (3.35), (3.40), and (3.49), if the semantic properties hold for it according to (3.20)–(3.22), (3.27)–(3.28), and (3.41)–(3.42), and if (3.50)–(3.52) hold for the enclosed views.

Concepts

We now formally define domain-specific concepts with a relational algebra. This includes the entities *Layer* and *Concept* in the domain metamodel and the association between them (cf. figure 3.5). We perform the formalization with the relation *Concept*, as illustrated in figure 3.7.

A concept is identified by a name and associated to a layer (3.14). The name of a concept represents the PK of the relation. The respective layer is indicated with its number as identifier. A layer can contain multiple concepts (3.14), and any concept belongs to one layer at least (3.14).

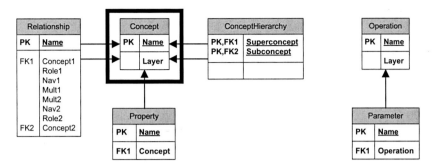

Figure 3.7: Concepts in the ER diagram

Definition 3 (*Concepts*)

Let Id_{La} be the set of layer identifiers, Id_{Co} the set of layer identifiers. Let $la \in Id_{La}, co \in Id_{Co}$.

The relation

$$Concept(Layer, Name) \subseteq Id_{La} \times Id_{Co} \tag{3.14}$$

specifies which concepts are modeled at which layers.

The set

$$Co := \{co | \exists la.(la, co) \in Concept\} \subseteq Id_{Co} \tag{3.15}$$

contains all modeled concepts independently of any layer.

The function

$$layersOfConcept : Id_{Co} \to \mathcal{P}(Id_{La})$$
$$co \mapsto \{la | (la, co) \in Concept\} \tag{3.16}$$

determines the layers where a particular concept is available.

The function

$$conceptsOfLayer : Id_{La} \to \mathcal{P}(Id_{Co})$$
$$la \mapsto \{co | (la, co) \in Concept\} \tag{3.17}$$

determines the available concepts at a particular layer.

The view

$$Concept_{la} := \{(la, co) | co \in conceptsOfLayer(la)\}$$
$$\subseteq Concept \tag{3.18}$$

shows the available concepts at a particular layer.

Concept Hierarchy

We now formally define the hierarchy of domain-specific concepts with a relational algebra. This includes the entity *Concept* in the domain metamodel and the association from and to itself (cf. figure 3.5). We perform the formalization with the relation *ConceptHierarchy*, as illustrated in figure 3.8.

The concept hierarchy realizes a single inheritance of concepts. A concept can have one superconcept (3.19, 3.21) and multiple subconcepts (3.19) that were modeled before (3.20). The respective concepts are referenced as FKs with their names. The combination of both references represent the PK of the relation. Furthermore, a concept must not be its own superconcept (3.22). This implies that a concept must not be its own subconcept and avoids cycles in the concept hierarchy.

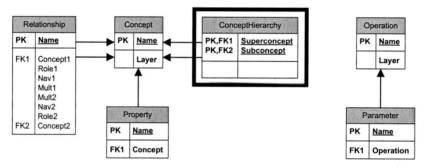

Figure 3.8: The concept hierarchy in the ER diagram

Definition 4 (*Concept hierarchy*)

Let Id_{La} be the set of layer identifiers, Id_{Co} the set of concept identifiers. Let Co be the set of modeled concepts. Let $la \in Id_{La}$, $co, co_1, co_2 \in Co$.

The relation

$$ConceptHierarchy(Superconcept, Subconcept) \subseteq Id_{Co} \times Id_{Co} \quad (3.19)$$

specifies which concepts are modeled as superconcept and which ones as subconcept.

Therefore, the following must hold:

$$(co_1, co_2) \in ConceptHierarchy \Rightarrow co_1, co_2 \in Co \tag{3.20}$$

$$(co_1, co) \in ConceptHierarchy \wedge (co_2, co) \in ConceptHierarchy$$
$$\Rightarrow co_1 = co_2 \tag{3.21}$$

$$co \notin superconceptsOfConcept(co) \tag{3.22}$$

The function

$$superconceptsOfConcept : Id_{Co} \rightarrow \mathcal{P}(Id_{Co})$$
$$co \mapsto \{co_2 | (co_2, co) \in ConceptHierarchy\}$$
$$\cup \bigcup_{(co_2, co) \in ConceptHierarchy} superconceptsOfConcept(co_2) \tag{3.23}$$

determines the superconcepts associated with a particular subconcept. This includes the direct superconcept, if existent, and recursively further superconcepts.

The function

$$subconceptsOfConcept : Id_{Co} \rightarrow \mathcal{P}(Id_{Co})$$
$$co \mapsto \{co_2 | (co, co_2) \in ConceptHierarchy\}$$
$$\cup \bigcup_{(co, co_2) \in ConceptHierarchy} subconceptsOfConcept(co_2) \tag{3.24}$$

determines the subconcepts associated with a particular superconcept. This includes the direct subconcepts and recursively further subconcepts.

The view

$$ConceptHierarchy_{la} := \{(co_1, co_2) | (co_1, co_2) \in ConceptHierarchy$$
$$\wedge co_1 \in conceptsOfLayer(la) \wedge co_2 \in conceptsOfLayer(la)\} \tag{3.25}$$
$$\subseteq ConceptHierarchy$$

shows the available concept hierarchy at a particular layer.

Properties

We now formally define properties of domain-specific concepts with a relational algebra. This includes the entities *Concept* and *Property* in the domain metamodel and the association between them (cf. figure 3.5). We perform the formalization with the relation *Property*, as illustrated in figure 3.9.

A property is identified by a name and associated to a concept (3.26) that was modeled before (3.27). The name of a property represents the PK of the relation. The respective concept is referenced as FK with its name. A concept can have multiple properties (3.26), and any property belongs to one concept (3.26, 3.28), so that different concepts must not share the same property.

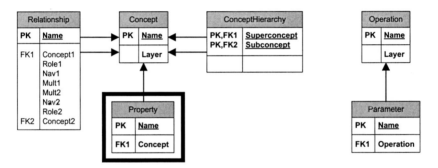

Figure 3.9: Properties in the ER diagram

Definition 5 (*Properties*)

Let Id_{La} be the set of layer identifiers, Id_{Co} the set of concept identifiers, Id_{Pr} the set of property identifiers. Let Co be the set of modeled concepts. Let $la \in Id_{La}$, $co, co_1, co_2 \in Id_{Co}$, $pr \in Id_{Pr}$.

The relation

$$Property(Concept, Name) \subseteq Id_{Co} \times Id_{Pr} \qquad (3.26)$$

specifies which properties are modeled with which concept. Therefore, the following must hold:

$$(co, pr) \in Property \Rightarrow co \in Co \qquad (3.27)$$
$$(co_1, pr) \in Property \wedge (co_2, pr) \in Property \Rightarrow co_1 = co_2 \qquad (3.28)$$

The set

$$Pr := \{pr|\exists co.(co, pr) \in Property\} \subseteq Id_{Pr} \tag{3.29}$$

contains all modeled properties independently of any layer.

The function

$$conceptOfProperty : Id_{Pr} \rightarrow Id_{Co}$$
$$pr \mapsto \begin{cases} co, & (co, pr) \in Property \\ undef, & otherwise \end{cases} \tag{3.30}$$

determines the concept associated with a particular property. Due to (3.28), the result is unique.

The function

$$propertiesOfConcept : Id_{Co} \rightarrow \mathcal{P}(Id_{Pr})$$
$$co \mapsto \{pr|(co, pr) \in Property\}$$
$$\cup \bigcup_{co_2 \in superconceptsOfConcept(co)} propertiesOfConcept(co_2) \tag{3.31}$$

determines the properties associated with a particular concept, including properties associated with superconcepts.

The function

$$layersOfProperty : Id_{Pr} \rightarrow \mathcal{P}(Id_{La})$$
$$pr \mapsto layersOfConcept(conceptOfProperty(pr)) \tag{3.32}$$

determines the layers where a particular property is available.

The function

$$propertiesOfLayer : Id_{La} \rightarrow \mathcal{P}(Id_{Pr})$$
$$la \mapsto \{pr|\exists co.(co, pr) \in Property \wedge la \in layersOfProperty(pr)\} \tag{3.33}$$

determines the available properties at a particular layer.

The view

$$Property_{la} := \{(co, pr)|(co, pr) \in Property$$
$$\wedge\ pr \in propertiesOfLayer(la)\} \subseteq Property \tag{3.34}$$

shows the available properties at a particular layer.

Operations

We now formally define domain-specific operations with a relational algebra. This includes the entities *Layer* and *Operation* in the domain metamodel and the association between them (cf. figure 3.5). We perform the formalization with the relation *Operation*, as illustrated in figure 3.10.

An operation is identified by a name and associated to a layer (3.35). The name of an operation represents the PK of the relation, and the respective layer is indicated with its number. A layer can contain multiple operations (3.35), and any operation belongs to one layer at least (3.35).

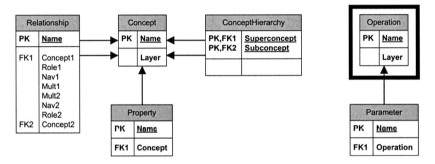

Figure 3.10: Operations in the ER diagram

Definition 6 (*Operations*)

Let Id_{La} be the set of layer identifiers, Id_{Op} the set of operation identifiers. Let $la \in Id_{La}, op \in Id_{Op}$.

The relation

$$Operation(Layer, Name) \subseteq Id_{La} \times Id_{Op} \tag{3.35}$$

specifies which operations are modeled at which layers.

The set

$$Op := \{op | \exists la.(la, op) \in Operation\} \subseteq Id_{Op} \tag{3.36}$$

contains all modeled operations independently of any layer.

The function

$$layersOfOperation : Id_{Op} \rightarrow \mathcal{P}(Id_{La})$$
$$op \mapsto \{la | (la, op) \in Operation\}$$

(3.37)

determines the layers where a particular operation is available.

The function

$$operationsOfLayer : Id_{La} \rightarrow \mathcal{P}(Id_{Op})$$
$$la \mapsto \{op | (la, op) \in Operation\}$$

(3.38)

determines the available operations at a particular layer.

The view

$$Operation_{la} := \{(la, op) | op \in operationsOfLayer(la)\}$$
$$\subseteq Operation$$

(3.39)

shows the available operations at a particular layer.

Parameters

We now formally define parameters of domain-specific operations with a relational algebra. This includes the entities *Operation* and *Parameter* in the domain metamodel and the association between them (cf. figure 3.5). We perform the formalization with the relation *Parameter*, as illustrated in figure 3.11.

A parameter is identified by a name and associated to an operation (3.40) that was modeled before (3.41). The name of a parameter represents the PK of the relation. The respective operation is referenced as FK with its name. An operation can have multiple parameters (3.40), and any parameter belongs to one operation (3.40, 3.42), so that different operations must not share the same parameter.

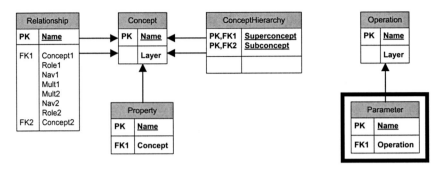

Figure 3.11: Parameters in the ER diagram

Definition 7 (*Parameters*)

Let Id_{La} be the set of layer identifiers, Id_{Op} the set of operation identifiers, Id_{Pa} the set of parameter identifiers. Let Op be the set of modeled operations. Let $la \in Id_{La}$, $op, op_1, op_2 \in Id_{Op}$, $pa \in Id_{Pa}$.

The relation

$$Parameter(Operation, Name) \subseteq Id_{Op} \times Id_{Pa} \tag{3.40}$$

specifies which parameters are modeled with which operation. Therefore, the following must hold:

$$(op, pa) \in Parameter \Rightarrow op \in Op \tag{3.41}$$

$$(op_1, pa) \in Parameter \wedge (op_2, pa) \in Parameter \Rightarrow op_1 = op_2 \tag{3.42}$$

The set

$$Pa := \{pa | \exists op.(op, pa) \in Parameter\} \subseteq Id_{Pa} \tag{3.43}$$

contains all modeled parameters independently of any layer.

The function

$$operationOfParameter : Id_{Pa} \to Id_{Op}$$

$$pa \mapsto \begin{cases} op, & (op, pa) \in Parameter \\ undef, & \text{otherwise} \end{cases} \tag{3.44}$$

determines the operation associated with a particular parameter. Due to (3.42), the result is unique.

The function

$$parametersOfOperation : Id_{Op} \rightarrow \mathcal{P}(Id_{Pa})$$
$$op \mapsto \{pa|(op, pa) \in Parameter\}$$

(3.45)

determines the parameters associated with a particular operation.

The function

$$layersOfParameter : Id_{Pa} \rightarrow \mathcal{P}(Id_{La})$$
$$pa \mapsto layersOfOperation(operationOfParameter(pa))$$

(3.46)

determines the layers where a particular parameter is available.

The function

$$parametersOfLayer : Id_{La} \rightarrow \mathcal{P}(Id_{Pa})$$
$$la \mapsto \{pa|\exists op.(op, pa) \in Parameter$$
$$\wedge la \in layersOfParameter(pa)\}$$

(3.47)

determines the available parameters at a particular layer.

The view

$$Parameter_{la} := \{(op, pa)|(op, pa) \in Parameter$$
$$\wedge pa \in parametersOfLayer(la)\} \subseteq Parameter$$

(3.48)

shows the available parameters at a particular layer.

Relationships

We now formally define relationships between domain-specific concepts with a relational algebra. This includes the entities *Concept*, *Relationship*, and *RelationshipEnd* in the domain metamodel and the association between them (cf. figure 3.5). We perform the formalization with the relation *Relationship*, as illustrated in figure 3.12.

A relationship is identified by its name (3.49). The name of a relationship represents the PK of the relation. Any relationship relates exactly two different concepts with each other (3.49,3.52) that were modeled before (3.50). The respective concepts are referenced as FKs with their names. Furthermore, the related concepts must share at least one common layer (3.51). Concepts can appear in multiple relationships (3.49).

The ends of a relationship specify the role of the respective concept within the relationship (3.49,3.52). The role is a textual value. Besides that, the ends also specify the navigability and multiplicity of the respective concept (3.49,3.52). A concept can be navigable or not in a relationship, and its multiplicity is a subset of the natural numbers, including zero and infinite.

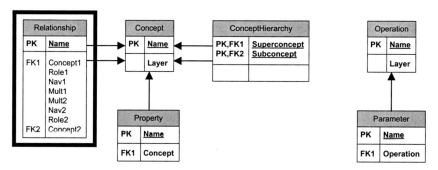

Figure 3.12: Relationships in the ER diagram

Definition 8 (*Relationships*)

Let Val_T be the set of textual values, Val_B the set of boolean values. Let Id_{La} be the set of layer identifiers, Id_{Co} the set of concept identifiers, Id_{Re} the set of relationship identifiers. Let Co be the set of modeled concepts. Let $co, co_1, co_2, co_3, co_4, sc, sc_2 \in Id_{Co}$, $re \in Id_{Re}$, $ro_1, ro_2, ro_3, ro_4 \in Val_T$, $na, na_1, na_2, na_3, na_4 \in Val_B$, $mu, mu_1, mu_2, mu_3, mu_4 \in \mathbb{N}_0$.

The relation

$$
\begin{aligned}
&Relationship(Concept1, Role1, Nav1, Mult1, Name, \\
&\quad Mult2, Nav2, Role2, Concept2) \\
&\quad \subseteq Id_{Co} \times Val_T \times Val_B \times \mathcal{P}(\mathbb{N}_0) \times Id_{Re} \\
&\quad\quad \times \mathcal{P}(\mathbb{N}_0) \times Val_B \times Val_T \times Id_{Co}
\end{aligned} \tag{3.49}
$$

specifies which concepts are in relationship with each other including their role, navigability, and multiplicity.

Therefore, the following must hold:

$$(co_1, ro_1, na_1, mu_1, re, mu_2, na_2, ro_2, co_2) \in Relationship$$
$$\Rightarrow co_1, co_2 \in Co \tag{3.50}$$

$$(co_1, ro_1, na_1, mu_1, re, mu_2, na_2, ro_2, co_2) \in Relationship$$
$$\Rightarrow layersOfConcept(co_1) \cap layersOfConcept(co_2) \neq \emptyset \tag{3.51}$$

$$(co_1, ro_1, na_1, mu_1, re, mu_2, na_2, ro_2, co_2) \in Relationship$$
$$\wedge (co_3, ro_3, na_3, mu_3, re, mu_4, na_4, ro_4, co_4) \in Relationship$$
$$\Rightarrow co_1 = co_3 \wedge co_2 = co_4 \wedge ro_1 = ro_3 \wedge ro_2 = ro_4 \tag{3.52}$$
$$\wedge na_1 = na_3 \wedge na_2 = na_4 \wedge mu_1 = mu_3 \wedge mu_2 = mu_4$$

The set

$$Re := \{re | \exists co_1, ro_1, na_1, mu_1, mu_2, na_2, ro_2, co_2.(co_1, ro_1,$$
$$na_1, mu_1, re, mu_2, na_2, ro_2, co_2) \in Relationship\} \subseteq Id_{Re} \tag{3.53}$$

contains all modeled relationships independently of any layer.

The function

$$conceptsOfRelationship : Id_{Re} \rightarrow Id_{Co} \times Id_{Co}$$

$$re \mapsto \begin{cases} (co_1, co_2), & \exists ro_1, na_1, mu_1, mu_2, ro_2, na_2. \\ & (co_1, ro_1, na_1, mu_1, re, mu_2, na_2, ro_2, co_2) \\ & \in Relationship \\ undef, & \text{otherwise} \end{cases} \tag{3.54}$$

determines the pair of concepts associated with a particular relationship. Due to (3.52), the result is unique.

The function

$$roles : Id_{Re} \rightarrow Val_T \times Val_T$$

$$re \mapsto \begin{cases} (ro_1, ro_2), & \exists co_1, na_1, mu_1, mu_2, na_2, co_2. \\ & (co_1, ro_1, na_1, mu_1, re, mu_2, na_2, ro_2, co_2) \\ & \in Relationship\} \\ undef, & \text{otherwise} \end{cases} \tag{3.55}$$

determines the pair of roles associated with a particular relationship. Due to (3.52), the result is unique.

The function

$$navigabilitiesOfRelationship : Id_{Re} \rightarrow Val_B \times Val_B$$

$$re \mapsto \begin{cases} (na_1, na_2), & \exists co_1, ro_1, mu_1, mu_2, ro_2, co_2. \\ & (co_1, ro_1, na_1, mu_1, re, mu_2, na_2, ro_2, co_2) \\ & \in Relationship\} \\ undef, & \text{otherwise} \end{cases} \tag{3.56}$$

determines the pair of navigabilities associated with a particular relationship. Due to (3.52), the result is unique.

The function

$$multiplicitiesOfRelationship : Id_{Re} \rightarrow \mathcal{P}(\mathbb{N}_0) \times \mathcal{P}(\mathbb{N}_0)$$

$$re \mapsto \begin{cases} (mu_1, mu_2), & \exists co_1, ro_1, na_1, na_2, ro_2, co_2. \\ & (co_1, ro_1, na_1, mu_1, re, mu_2, na_2, ro_2, co_2) \\ & \in Relationship\} \\ undef, & \text{otherwise} \end{cases} \tag{3.57}$$

determines the pair of multiplicities associated with a particular relationship. Due to (3.52), the result is unique.

The function

$$relationshipsOfConcept : Id_{Co} \rightarrow \mathcal{P}(Id_{Re})$$
$$co \mapsto \{re | \exists ro_1, na_1, mu_1, mu_2, na_2, ro_2, co_2.$$
$$(co, ro_1, na_1, mu_1, re, mu_2, na_2, ro_2, co_2) \in Relationship\}$$
$$\cup \{re | \exists co_1, ro_1, na_1, mu_1, mu_2, na_2, ro_2. \tag{3.58}$$
$$(co_1, ro_1, na_1, mu_1, re, mu_2, na_2, ro_2, co) \in Relationship\}$$
$$\cup \bigcup_{co_2 \in superconceptsOfConcept(co)} relationshipsOfConcept(co_2)$$

determines the relationship identifiers associated with a particular concept, including relationships associated with superconcepts.

The function

$$layersOfRelationship : Id_{Re} \rightarrow \mathcal{P}(Id_{La})$$
$$re \mapsto layersOfConcept(co_1) \cap layersOfConcept(co_2) \tag{3.59}$$
$$\text{with } (co_1, co_2) = conceptsOfRelationship(re)$$

determines the layers where a particular relationship is available.

The function

$$relationshipsOfLayer : Id_{La} \rightarrow \mathcal{P}(Id_{Re})$$
$$la \mapsto \{re | \exists co_1, ro_1, na_1, mu_1, mu_2, na_2, ro_2, co_2.$$
$$(la, co_1) \in Concept \wedge (la, co_2) \in Concept$$
$$\wedge (co_1, ro_1, na_1, mu_1, re, mu_2, na_2, ro_2, co_2) \in Relationship\}$$

(3.60)

determines the available relationships at a particular layer.

The view

$$Relationship_{la} := \{(co_1, ro_1, na_1, mu_1, re, mu_2, na_2, ro_2, co_2) |$$
$$(co_1, ro_1, na_1, mu_1, re, mu_2, na_2, ro_2, co_2) \in Relationship$$
$$\wedge re \in relationshipsOfLayer(la)\} \subseteq Relationship$$

(3.61)

shows the available relationships at a particular layer.

The function

$$directlyRelatedConcepts : Id_{Co} \rightarrow \mathcal{P}(Id_{Co})$$
$$co \mapsto \{co_2 | \exists re, sc, sc_2.re \in Re$$
$$\wedge sc \in \{co\} \cup superconceptsOfConcept(co)$$
$$\wedge sc_2 \in \{co_2\} \cup superconceptsOfConcept(co_2)$$
$$\wedge (conceptsOfRelationship(re) = (sc, sc_2)$$
$$\vee conceptsOfRelationship(re) = (sc_2, sc))\}$$

(3.62)

determines the concepts that are directly related to another concept. A concept is directly related to another concept if a relationship is modeled between the two concepts. This relationship can also be modeled between one of the two concepts and a superconcept of the other one, or between a superconcept of one of the two concepts and a superconcept of the other one. This bears in mind that relationships are inherited from superconcepts to subconcepts, i.e. it makes no difference for a concept whether a relationship is modeled from the concept itself or from any of its superconcepts.

The function

$$relatedConcepts : Id_{Co} \rightarrow \mathcal{P}(Id_{Co})$$
$$co \mapsto \{co\} \cup \bigcup_{co_2 \in directlyRelatedConcepts(co)} relatedConcepts(co_2)$$

(3.63)

determines the concepts that are related to another concept. A concept is related to itself and to any concept to which it is directly related over a chain of relationships.

The function

$$directlyAccessibleConcepts : Id_{Co} \rightarrow \mathcal{P}(Id_{Co})$$
$$co \mapsto \{co_2 | \exists re, sc, sc_2.re \in Re$$
$$\wedge sc \in \{co\} \cup superconceptsOfConcept(co)$$
$$\wedge sc_2 \in \{co_2\} \cup superconceptsOfConcept(co_2)$$
$$\wedge (conceptsOfRelationship(re) = (sc, sc_2)$$
$$\wedge \exists mu.multiplicitiesOfRelationship(re) = (mu, 1)$$
$$\wedge \exists na.navigabilitiesOfRelationship(re) = (na, true)$$
$$\vee conceptsOfRelationship(re) = (sc_2, sc)$$
$$\wedge \exists mu.multiplicitiesOfRelationship(re) = (1, mu)\}$$
$$\wedge \exists na.navigabilitiesOfRelationship(re) = (true, na))\}$$

(3.64)

determines the concepts that are directly accessible from another concept. These are determined as follows. For a concept to be directly accessible from another concept, a relationship must be modeled between the two concepts at first. This relationship can also be modeled between one of the two concepts and a superconcept of the other one, or between a superconcept of one of the two concepts and a superconcept of the other one. This again bears in mind that relationships are inherited from superconcepts to subconcepts, i.e. it makes no difference for a concept whether a relationship is modeled from the concept itself or from any of its superconcepts. Secondly, for a concept to be directly accessible from another concept, the multiplicity of the relationship at the end of the accessible concept must be 1. Higher multiplicities result in a set of concepts at this end of the relationship and thus prohibit the unique referring to a single concept out of this set of concepts. Thirdly, the accessible concept must be navigable at its end of the relationship.

The function

$$accessibleConcepts : Id_{Co} \rightarrow \mathcal{P}(Id_{Co})$$
$$co \mapsto \{co\} \cup \bigcup_{co_2 \in directlyAccessibleConcepts(co)} accessibleConcepts(co_2)$$

(3.65)

determines the concepts that are accessible from another concept. A concept is accessible from itself and from any concept from which it is directly accessible over a chain of relationships.

3.3.3 Example

In the following, we present a domain model that specifies the domain of the example scenario provided in section 3.1. We model the domain at two abstraction layers. The first layer contains the concepts of the scenario description, whereas the second layer provides a technical representation of the first layer. Figure 3.13 shows an overview of the domain model with a concrete syntax based on Unified Modeling Language (UML). Tables 3.1 to 3.6 provide the formal representation of the domain model.

At the first abstraction layer, we represent a cell phone with the concept *cellPhone*. The property *cpImei* of this concept serves as a unique identifier for cell phones. Then, we represent a variation of signal quality with the concept *signalQuality*. The property *sqCellPhoneImei* of the concept refers to the cell phone for which a variation of signal quality was detected, and the properties *sqOldValue* and *sqNewValue* contain the value before and after the variation. We also model two operations *increasePower* and *decreasePower* to adjust the transmission power of the antenna.

Due to the technical focus of the second layer, different terminology is used to describe the same domain. Here, a cell phones is more generally regarded as a *device*, as modeled with the respective concept. A device is uniquely identified with the property *dId*. A variation of signal quality is represented by the concept *intensityChange*. The property *icDeviceId* of the concept again refers to the respective device, and the properties *icOldValue* and *icNewValue* contain the intensity of the signal before and after the change. The two operations to increase and decrease the transmission power are now merged into one operation *changeTXP*. The parameter *ctChangeValue* is used to specify how to change the TXP of the antenna.

As the example scenario is rather simple, the domain model does not contain any concept hierarchy or relationship. However, concept hierarchies and relationships appear in the case studies in chapter 5.

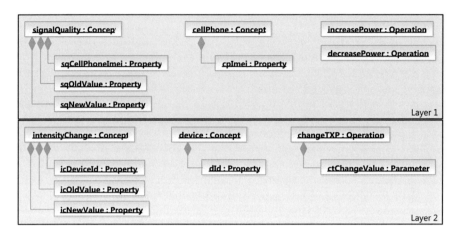

Figure 3.13: Example domain model (layers 1 and 2)

Table 3.1: Relation *Concept*

Layer	Name
1	signalQuality
1	cellPhone
2	intensityChange
2	device

Table 3.2: Relation *ConceptHierarchy*

Superconcept	Subconcept

Table 3.4: Relation *Operation*

Layer	Name
1	increasePower
1	decreasePower
2	changePower

Table 3.3: Relation *Property*

Concept	Name
signalQuality	sqCellPhoneImei
signalQuality	sqOldValue
signalQuality	sqNewValue
cellPhone	cpImei
intensityChange	icDeviceId
intensityChange	icOldValue
intensityChange	icNewValue
device	dId

Table 3.5: Relation *Parameter*

Operation	Name
changePower	cpChangeValue

Table 3.6: Relation *Relationship*

Concept1	Role1	Nav1	Mult1	Name	Mult2	Nav2	Role2	Concept2

3.4 Policy Modeling

In the same way as expert groups have a particular view of the domain, they also have a particular view of the policies that control a system in that domain. An application expert e.g. uses a different terminology to express a policy than a system administrator does for the same policy. Also, an application policy can be represented by several technical policies at a lower abstraction layer.

We cover structural information about the policies with the policy model, as illustrated in figure 3.14. The policy model specifies which policies are used to manage the system and which structure they have. Policy experts specify the policy model at design time and can change it at runtime whenever the system behavior should be changed. As with domain modeling, the policy model is an instance of the policy metamodel and offers a particular view at any layer. Views filter the part of the policy model that is visible at the respective layer. Two layers i and j are shown exemplarily in figure 3.14.

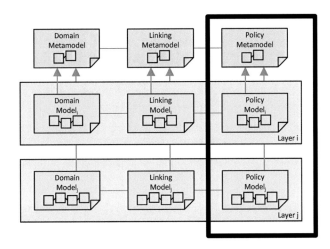

Figure 3.14: Policy modeling

In the following sections, we describe how we use the policy metamodel and the policy model to specify policies. A relational algebra formally defines the policy metamodel, i.e. the abstract syntax of the policy model.

3.4.1 Policy Metamodel

An important aspect of policy development is an initial modeling and an automated code generation from the models. Policies should be specified independently of any policy language at first. Afterwards, code for a particular language should be generated in an automated way. A policy language usually provides a textual notation for policies and can be interpreted by an execution engine. Typically, such engines provide means to cope with priorities and conflict resolution, and this reduces development complexity.

Unfortunately, no policy language is general and powerful enough to meet the requirements of any policy-based system in any domain. When developing a policy-based system, a decision for a policy language has to be made at some point. The decision for an existing language heavily depends on the specific problem to solve and on technical details of the respective system. Different policy languages are available at the CIM, PIM, and PSM levels for different purposes. Developing an individual policy language and engine from scratch is an extensive task and should usually be omitted. Instead, an existing policy language should be used whenever possible.

For this reason, we analyzed well-known policy languages including PonderTalk [102], KAoS [160], and Rei [161]. Based on this analysis, we integrated common concepts into a policy metamodel that covers important means to specify ECA policies. The focus on common concepts ensures an automated transformation of policy models into executable code. The resulting metamodel is shown in figure 3.15. It offers a flexible number of abstraction layers and allows to specify policies independently of any language and domain. Domain-specific aspects are modeled with the linking model later.

The policy metamodel represents the abstract syntax of the policies, i.e. it defines the structure of the policy model. It contains the following entities. *Policies* have a name and are active or not. They are assigned to *layers* to specify at which levels of abstraction they are available. Each layer contains an arbitrary number of policies, and each policy is assigned to one layer at least. A policy has named *events* assigned, and a policy optionally has a named *condition* assigned. A condition is one out of the following boolean expressions. A *binary expression* performs a comparison between two arguments and is one out of *equal, greater, greater or equal, lower,* and *lower or equal.* Section 3.5.1 defines the possible content of these arguments. A *negation expression* negates another expression. An *and expression* combines two other expressions as conjunction, an *or expression* as disjunction. A special type of expression is the *operational expression*, which uses the return value of an invoked operation as the result of the expression. Finally, a policy has named *actions* assigned. The sequence of actions within that policy is specified by the *action informations*, assigning an execution number to each action.

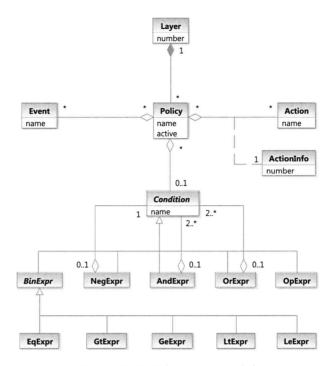

Figure 3.15: Policy metamodel

Again, a layer is unambiguously identified by its number and this number is used to denote the layer. Therefore, the number of a layer is unique. Layers are ordered in a top-down way, starting with *1*. $la_1 < la_2$ means that layer la_1 contains less technical details and thus represents a higher level of abstraction than layer la_2, and vice versa. Policies, events, conditions, and actions are unambiguously identified by their name and their name is used to denote them. Therefore, their names are unique.

3.4.2 Policy Model

The policy model is used to specify policies and their structure, i.e. which events, conditions, and actions are modeled with which policies. Furthermore, policies are assigned to various abstraction layers. Various restrictions must hold on the policy model as the policy model is an instance of the policy metamodel.

In the following, we formally define the abstract syntax of the policy model with a relational algebra. Figure 3.16 shows an overview of the respective relations as an ER diagram.

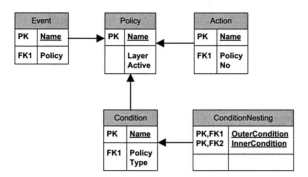

Figure 3.16: ER diagram of the policy model

Definition 9 (*Policy model*)

Let *Policy* be a relation of policies, *Event* a relation of events, *Condition* a relation of conditions, *ConditionNesting* a relation of condition nesting, *Action* a relation of actions.

We define the policy model

$$PM := (Policy, Event, Condition, \qquad\qquad (3.66)$$
$$ConditionNesting, Action)$$

as a quintuple of relations that specifies which policies, events, conditions, and actions are modeled. A policy model is valid if the model is well-formed according to (3.68), (3.75), (3.84), (3.95), and (3.105), and if the semantic properties hold for it according to (3.69), (3.76), (3.85)–(3.86), (3.96)–(3.101), and (3.106)–(3.107).

Let furthermore $Policy_{la}$ be a view of *Policy*, $Event_{la}$ a view of *Event*, $Condition_{la}$ a view of *Condition*, $ConditionNesting_{la}$ a view of *ConditionNesting*, $Action_{la}$ a view of *Action* at layer *la*.

We define the view of the policy model *PM* at layer *la*

$$PM_{la} := (Policy_{la}, Event_{la}, Condition_{la}, \qquad\qquad (3.67)$$
$$ConditionNesting_{la}, Action_{la})$$

as a quintuple of views that specifies which policies, events, conditions, and actions are modeled at a particular layer. A view of a policy model is valid if the view is well-formed according to (3.68), (3.75), (3.84), (3.95), and (3.105), and if the semantic properties hold for it according to (3.69), (3.76), (3.85)–(3.86), (3.96)–(3.101), and (3.106)–(3.107).

Policies

We now formally define policies with a relational algebra. This includes the entities *Layer* and *Policy* in the policy metamodel and the association between them (cf. figure 3.15). We perform the formalization with the relation *Policy*, as illustrated in figure 3.17.

A policy is identified by a name and associated to one layer (3.68). The name of a policy represents the PK of the relation. The respective layer is indicated with its number. If a policy should be used at multiple layers, it must be modeled at each layer with a different name. A layer can contain multiple policies (3.68), and any policy belongs to one layer (3.68,3.69), so that different layers must not share the same policy. Furthermore, any policy is active or inactive (3.68).

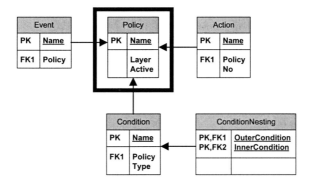

Figure 3.17: Policies in the ER diagram

Definition 10 (*Policies*)

Let Val_B be the set of boolean values. Let Id_{La} be the set of layer identifiers, Id_{Po} the set of policy identifiers. Let $act, act_1, act_2 \in Val_B$, $la, la_1, la_2 \in Id_{La}$, $po \in Id_{Po}$.

The relation

$$Policy(Layer, Name, Active) \subseteq Id_{La} \times Id_{Po} \times Val_B \tag{3.68}$$

specifies which policies are modeled at which layer and which policies are active.

Therefore, the following must hold:

$$(la_1, po, act_1) \in Policy \land (la_2, po, act_2) \in Policy \qquad (3.69)$$
$$\Rightarrow la_1 = la_2 \land act_1 = act_2$$

The set

$$Po := \{po | \exists la, act.(la, po, act) \in Policy\} \subseteq Id_{Po} \qquad (3.70)$$

contains all modeled policies independently of any layer.

The function

$$activityOfPolicy : Id_{Po} \rightarrow Val_B$$
$$po \mapsto \begin{cases} act, & \exists la.(la, po, act) \in Policy \\ undef, & \text{otherwise} \end{cases} \qquad (3.71)$$

determines whether a particular policy is active or not.

The function

$$layerOfPolicy : Id_{Po} \rightarrow Id_{La}$$
$$po \mapsto \begin{cases} la, & \exists act.(la, po, act) \in Policy \\ undef, & \text{otherwise} \end{cases} \qquad (3.72)$$

determines the layer where a particular policy is used. Due to (3.69), the result is unique.

The function

$$policiesOfLayer : Id_{La} \rightarrow \mathcal{P}(Id_{Po})$$
$$la \mapsto \{po | \exists act.(la, po, act) \in Policy\} \qquad (3.73)$$

determines the used policies at a particular layer.

The view

$$Policy_{la} := \{(la, po, act) | (la, po, act) \in Policy$$
$$\land po \in policiesOfLayer(la)\} \subseteq Policy \qquad (3.74)$$

shows the used policies at a particular layer.

Events

We now formally define policy events with a relational algebra. This includes the entities *Policy* and *Event* in the policy metamodel and the association between them (cf. figure 3.15). We perform the formalization with the relation *Event*, as illustrated in figure 3.18.

An event is identified by a name and can belong to a set of policies (3.75) that were modeled before (3.76). The name of an event represents the PK of the relation. The respective policy is referenced as FK with its name. *undef* as reference indicates that the event does not belong to a policy. A policy can have multiple events (3.75), and any event belongs to at least one policy (3.75).

As mentioned before, the policy model only specifies the structure of the policies, but not their domain-specific content. Here, it only specifies which events are used, but not how they are represented in the specific domain. This information is specified in the linking model, as shown in section 3.5.1.

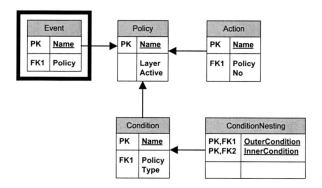

Figure 3.18: Events in the ER diagram

Definition 11 (*Events*)

Let Val_B be the set of boolean values. Let Id_{La} be the set of layer identifiers, Id_{Po} the set of policy identifiers, Id_{Ev} the set of event identifiers. Let Po be the set of modeled policies. Let $act \in Val_B$, $la \in Id_{La}$, $po \in Id_{Po}$, $ev \in Id_{Ev}$.

The relation

$$Event(Policy, Name) \subseteq Id_{Po} \cup \{undef\} \times Id_{Ev} \qquad (3.75)$$

specifies which events are modeled with which policies.

Therefore, the following must hold:

$$(po, ev) \in Event \Rightarrow po \in Po \vee po = undef \tag{3.76}$$

The set

$$Ev := \{ev | \exists po.(po, ev) \in Event\} \subseteq Id_{Ev} \tag{3.77}$$

contains all modeled events independently of any layer.

The function

$$\begin{aligned}
&policiesOfEvent : Id_{Ev} \rightarrow \mathcal{P}(Id_{Po})\\
&ev \mapsto \{po|(po, ev) \in Event\} \cap Id_{Po}
\end{aligned} \tag{3.78}$$

determines the policies to which a particular event belongs. *undef* references are excluded from the result by the intersection with the set of policy identifiers.

The function

$$\begin{aligned}
&eventsOfPolicy : Id_{Po} \rightarrow \mathcal{P}(Id_{Ev})\\
&po \mapsto \{ev|(po, ev) \in Event\}
\end{aligned} \tag{3.79}$$

determines the events that belong to a particular policy.

The function

$$\begin{aligned}
&layersOfEvent : Id_{Ev} \rightarrow \mathcal{P}(Id_{La})\\
&ev \mapsto \bigcup_{po \in policiesOfEvent(ev)} layerOfPolicy(po)
\end{aligned} \tag{3.80}$$

determines the layers where a particular event is used.

The function

$$\begin{aligned}
&eventsOfLayer : Id_{La} \rightarrow \mathcal{P}(Id_{Ev})\\
&la \mapsto \{ev|\exists po.(po, ev) \in Event \wedge la \in layersOfEvent(ev)\}
\end{aligned} \tag{3.81}$$

determines the used events at a particular layer.

The view

$$\begin{aligned}
Event_{la} := \{&(po, ev)|(po, ev) \in Event\\
&\wedge po \in policiesOfLayer(la) \wedge ev \in eventsOfLayer(la)\}\\
&\subseteq Event
\end{aligned} \tag{3.82}$$

shows the used events at a particular layer.

Conditions

We now formally define policy conditions with a relational algebra. This includes the entities *Policy, Condition, BinExpr, EqExpr, GtExpr, GeExpr, LtExpr, LeExpr, NeqExpr, AndExpr, OrExpr,* and *OpExpr* in the policy metamodel and the associations between them (cf. figure 3.15). We perform the formalization with the relation *Condition*, as illustrated in figure 3.19.

A condition is identified by a name and can belong to a policy (3.84) that was modeled before (3.85). The name of a condition represents the PK of the relation. The respective policy is referenced as FK with its name. *undef* as reference indicates that the condition does not belong to a policy. A policy can have one condition at most (3.84,3.86), and any condition belongs to an arbitrary number of policies (3.84). If a condition does not belong to any policy, the value *undef* is used

A condition is a boolean expression that is evaluated to *true* or *false*. For this purpose, any condition has a type (3.84), which indicates the operator to evaluate the condition. According to the condition type, the operator is one out of the following (3.83):

- Comparison of two arguments: equal to (*eq*), greater than (*gt*), greater than or equal to (*ge*), less than (*lt*), less than or equal to (*le*)

- Connective of other conditions: negation (*not*), conjunction (*and*), disjunction (*or*)

- Invoking of an operation with a boolean return value: operational (*op*)

The policy model only specifies which conditions are used, but not which arguments are compared or which operations are invoked in these conditions. This information is specified in the linking model, as shown in section 3.5.1.

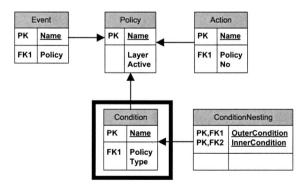

Figure 3.19: Conditions in the ER diagram

Definition 12 (*Conditions*)

The set

$$Type_{Cd} := \{eq, gt, ge, lt, le, not, and, or, op\} \subseteq Val \tag{3.83}$$

contains the different types of condition.

Let furthermore Val_B be the set of boolean values. Let Id_{La} be the set of layer identifiers, Id_{Po} the set of policy identifiers, Id_{Cd} the set of condition identifiers. Let Po be the set of modeled policies. Let $act \in Val_B$, $la \in Id_{La}$, $po \in Id_{Po} \cup \{undef\}$, $cd, cd_1, cd_2 \in Id_{Cd}$, $ty, ty_1, ty_2 \in Type_{Cd}$.

The relation

$$\begin{aligned} &Condition(Policy, Name, Type) \\ &\quad \subseteq Id_{Po} \cup \{undef\} \times Id_{Cd} \times Type_{Cd} \end{aligned} \tag{3.84}$$

specifies which conditions are modeled with which policies and which type they have. Therefore, the following must hold:

$$(po, cd, ty) \in Condition \Rightarrow po \in Po \lor po = undef \tag{3.85}$$

$$\begin{aligned} &(po, cd_1, ty_1) \in Condition \land (po, cd_2, ty_2) \in Condition \\ &\quad \Rightarrow cd_1 = cd_2 \land ty_1 = ty_2 \end{aligned} \tag{3.86}$$

The set

$$Cd := \{cd | \exists po, ty.(po, cd, ty) \in Condition\} \subseteq Id_{Cd} \tag{3.87}$$

contains all modeled conditions independently of any layer.

The function

$$\begin{aligned} &typeOfCondition : Id_{Cd} \rightarrow Type_{Cd} \\ &cd \mapsto \begin{cases} ty, & \exists po.(po, cd, ty) \in Condition \\ undef, & \text{otherwise} \end{cases} \end{aligned} \tag{3.88}$$

determines the type of a particular condition.

The function

$$\begin{aligned} &policiesOfCondition : Id_{Cd} \rightarrow \mathcal{P}(Id_{Po}) \\ &cd \mapsto \{po | \exists ty.(po, cd, ty) \in Condition\} \cap Id_{Po} \\ &\quad \cup \bigcup_{cd_2 \in outerConditionsOfCondition(cd)} policiesOfCondition(cd_2) \end{aligned} \tag{3.89}$$

determines the policies to which a particular condition belongs, directly as condition or indirectly as a subcondition. The function $outer-ConditionsOfCondition$ is provided in 3.102. $undef$ references are excluded from the result by the intersection with the set of policy identifiers.

The function

$$conditionOfPolicy : Id_{Po} \rightarrow Id_{Cd}$$

$$po \mapsto \begin{cases} cd, & \exists ty.(po, cd, ty) \in Condition \\ undef, & \text{otherwise} \end{cases} \qquad (3.90)$$

determines the condition that directly belongs to a particular policy. Due to 3.86, the result is unique.

The function

$$conditionsOfPolicy : Id_{Po} \rightarrow \mathcal{P}(Id_{Cd})$$

$$po \mapsto conditionOfPolicy(po) \qquad (3.91)$$
$$\cup innerConditionsOfCondition(conditionOfPolicy(po))$$

determines the conditions that belong to a particular policy, directly as condition or indirectly as a subcondition. The function $innerConditionsOfCondition$ is provided in 3.103.

The function

$$layersOfCondition : Id_{Cd} \rightarrow \mathcal{P}(Id_{La})$$

$$cd \mapsto \bigcup_{po \in policiesOfCondition(cd)} layerOfPolicy(po) \qquad (3.92)$$

determines the layers where a particular condition is used.

The function

$$conditionsOfLayer : Id_{La} \rightarrow \mathcal{P}(Id_{Cd})$$

$$la \mapsto \{cd | \exists po, ty.(po, cd, ty) \in Condition \qquad (3.93)$$
$$\wedge la \in layersOfCondition(cd)\}$$

determines the used conditions at a particular layer.

The view

$$Condition_{la} := \{(po, cd, ty) | (po, cd, ty) \in Condition$$
$$\wedge po \in policiesOfLayer(la) \cup \{undef\} \qquad (3.94)$$
$$\wedge cd \in conditionsOfLayer(la)\} \subseteq Condition$$

shows the used conditions at a particular layer.

Condition Nesting

We now formally define the nesting of policy conditions with a relational algebra. This includes the entity *Condition* in the policy metamodel and the association from and to itself (cf. figure 3.15). We perform the formalization with the relation *ConditionNesting*, as illustrated in figure 3.20.

The condition nesting realizes a single inheritance of conditions, and this allows to combine multiple inner conditions to more complex outer conditions. The usage of a condition as inner condition means that it does not belong to a policy in this context (3.99). A condition can be contained in one outer condition (3.95,3.97) and can contain multiple inner conditions (3.95) that were modeled before (3.96). The respective inner and outer conditions are referenced as FKs with their names. The combination of both references represent the PK of the relation. Furthermore, a condition must not be contained in itself (3.98). This implies that a condition must not contain itself and avoids cycles in the condition nesting.

Conditions of a binary (*eq, lt, le, gt, ge*) and operational (*oe*) type are atomic and must not contain any inner conditions (3.100). A negating condition (type *not*) must contain the negated condition as inner condition (3.101). A conjuncting or disjuncting condition (types *and, or*) contains an arbitrary number of inner conditions.

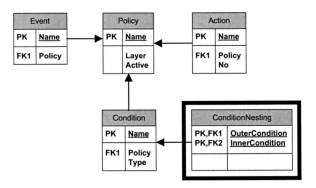

Figure 3.20: Condition nesting in the ER diagram

Definition 13 (*Condition nesting*)

Let Id_{La} be the set of layer identifiers, Id_{Cd} the set of condition identifiers. Let Po the set of modeled policies, Cd the set of modeled conditions, $Type_{Cd}$ the set of condition types. Let $la \in Id_{La}$, $po \in Po$, $cd, cd_1, cd_2 \in Cd$, $ty \in Type_{Cd}$.

The relation

$$ConditionNesting(OuterCondition, InnerCondition)$$
$$\subseteq Id_{Cd} \times Id_{Cd} \tag{3.95}$$

specifies which conditions are modeled as outer condition and which ones as inner condition. Therefore, the following must hold:

$$(cd_2, cd_1) \in ConditionNesting \Rightarrow cd_2, cd_1 \in Cd \tag{3.96}$$

$$(cd_1, cd) \in ConditionNesting \wedge (cd_2, cd) \in ConditionNesting$$
$$\Rightarrow cd_1 = cd_2 \tag{3.97}$$

$$(cd_2, cd_1) \in ConditionNesting$$
$$\Rightarrow cd_1 \notin outerConditionsOfCondition(cd_1) \tag{3.98}$$

$$(cd_2, cd_1) \in ConditionNesting$$
$$\Rightarrow \exists ty.(undef, cd_1, ty) \in Condition \tag{3.99}$$

$$(po, cd, ty) \in Condition \wedge ty \in \{eq, gt, ge, lt, le, op\}$$
$$\Leftrightarrow |innerConditionsOfCondition(cd)| = 0 \tag{3.100}$$

$$(po, cd, ty) \in Condition \wedge ty \in \{not\}$$
$$\Leftrightarrow |innerConditionsOfCondition(cd)| = 1 \tag{3.101}$$

The function

$$outerConditionsOfCondition : Id_{Cd} \rightarrow \mathcal{P}(Id_{Cd})$$
$$cd \mapsto \{cd_2|(cd_2, cd) \in ConditionNesting\}$$
$$\cup \bigcup_{(cd_2, cd) \in ConditionNesting} outerConditionsOfCondition(cd_2) \tag{3.102}$$

determines the outer conditions that contain a particular inner condition. This includes the direct outer condition, if existent, and recursively further outer conditions.

The function

$$innerConditionsOfCondition : Id_{Cd} \rightarrow \mathcal{P}(Id_{Cd})$$
$$cd \mapsto \{cd_2|(cd, cd_2) \in ConditionNesting\}$$
$$\cup \bigcup_{(cd, cd_2) \in ConditionNesting} innerConditionsOfCondition(cd_2) \tag{3.103}$$

determines the inner conditions that are contained in a particular outer condition. This includes the direct inner conditions and recursively further inner conditions.

The view

$$ConditionNesting_{la} := \{(cd_1, cd_2) | (cd_1, cd_2) \in ConditionNesting$$
$$\wedge\, cd_1 \in conditionsOfLayer(la) \wedge cd_2 \in conditionsOfLayer(la)\}$$
$$\subseteq ConditionNesting$$

(3.104)

shows the available condition nesting at a particular layer.

Actions

We now formally define policy actions with a relational algebra. This includes the entities *Policy* and *Action* in the policy metamodel and the association between them (cf. figure 3.15). We perform the formalization with the relation *Action*, as illustrated in figure 3.21.

An action is identified by a name and can belong to a set of policies (3.105) that were modeled before (3.106). The name of an action represents the PK of the relation. The respective policy is referenced as FK with its name. *undef* as reference indicates that the event does not belong to a policy. A policy can have multiple actions (3.105), and any action belongs to at least one policy (3.105). Furthermore, multiple actions are executed in a particular sequence, as indicated by an execution number for each action (3.105). Any execution number must only be used once within the same policy (3.107).

The policy model only specifies which actions are used, but not which operations are invoked in these actions. This information is specified in the linking model, as shown in section 3.5.1.

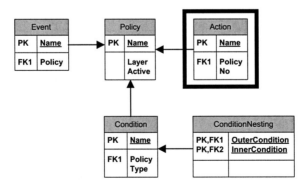

Figure 3.21: Actions in the ER diagram

Definition 14 (*Actions*)

Let Val_B be the set of boolean values. Let Id_{La} be the set of layer identifiers, Id_{Po} the set of policy identifiers, Id_{Ac} the set of action identifiers. Let Po be the set of modeled policies. Let $act \in Val_B$, $la \in Id_{La}$, $po \in Id_{Po}$, $ac, ac_1, ac_2 \in Id_{Ac}$, $no \in \mathbb{N}$.

The relation

$$Action(Policy, Name, No)$$
$$\subseteq Id_{Po} \cup \{undef\} \times Id_{Ac} \times \mathbb{N} \cup \{undef\} \tag{3.105}$$

specifies which actions are modeled with which policies and which execution number they have. Therefore, the following must hold:

$$(po, ac, no) \in Action \Rightarrow po \in Po \vee po = undef \tag{3.106}$$
$$(po, ac_1, no) \in Action \wedge (po, ac_2, no) \in Action \Rightarrow ac_1 = ac_2 \tag{3.107}$$

The set

$$Ac := \{ac | \exists po, no.(po, ac, no) \in Action\} \subseteq Id_{Ac} \tag{3.108}$$

contains all modeled actions independently of any layer.

The function

$$policiesOfAction : Id_{Ac} \rightarrow \mathcal{P}(Id_{Po})$$
$$ac \mapsto \{po | \exists no.(po, ac, no) \in Action\} \cap Id_{Po} \tag{3.109}$$

determines the policies to which a particular action belongs. *undef* references are excluded from the result by the intersection with the set of policy identifiers.

The function

$$actionsOfPolicy : Id_{Po} \rightarrow \mathcal{P}(Id_{Ac})$$
$$po \mapsto \{ac | \exists no.(po, ac, no) \in Action\} \tag{3.110}$$

determines the actions that belong to a particular policy.

The function

$$layersOfAction : Id_{Ac} \rightarrow \mathcal{P}(Id_{La})$$
$$ac \mapsto \bigcup_{po \in policiesOfAction(ac)} layerOfPolicy(po) \tag{3.111}$$

determines the layers where a particular action is used.

The function

$$actionsOfLayer : Id_{La} \to \mathcal{P}(Id_{Ac})$$
$$la \mapsto \{ac | \exists po, no.(po, ac, no) \in Action \tag{3.112}$$
$$\wedge\, la \in layersOfAction(ac)\}$$

determines the used actions at a particular layer.

The view

$$Action_{la} := \{(po, ac, no) | (po, ac, no) \in Action$$
$$\wedge\, po \in policiesOfLayer(la) \wedge ac \in actionsOfLayer(la)\} \tag{3.113}$$
$$\subseteq Action$$

shows the used actions at a particular layer.

3.4.3 Example

In the following, we present a policy model that specifies the policies of the example scenario provided in section 3.1. We model the policies at the upper abstraction layer only. The policies at the lower layer are generated later by an automated refinement of the upper layer, as shown in section 4.2.4. Figure 3.22 shows an overview of the policy model with a UML-based concrete syntax. Tables 3.7 to 3.11 provide the formal representation of the policy model.

At first, we model a policy *lowQuality* for dealing with signal quality that falls below a critical value. For this purpose, the policy is triggerd by an event *lqEvent* and analyzes the variation of signal quality with a condition *lqCondition*. This condition is of type *lt* in order to perform a lower than comparison later. If the contition is met, the policy executes its action *lqAction*. The action is modeled with an execution number of *1* as it is the only action of the policy. In the same way, we model a policy *highQuality* for dealing with signal quality that goes beyond another critical value. The event, condition, and action of this policy are called *hqEvent*, *hqCondition*, and *hqAction*. This condition is of type *gt* in order to perform a greater than comparison later. The action is again modeled with an execution number of *1*.

The details of the events, conditions, and actions are specified in the linking model in section 3.5.3. As the example scenario is rather simple, the policy model does not contain any condition nesting. However, condition nestings appear in the case studies in chapter 5.

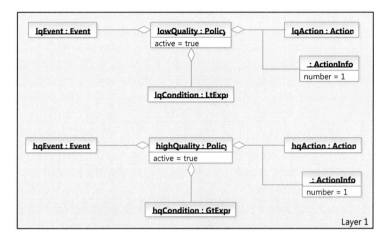

Figure 3.22: Example policy model (layer 1)

Table 3.7: View *Policy*$_1$

Layer	Name	Active
1	lowQuality	true
1	highQuality	true

Table 3.8: View *Event*$_1$

Policy	Name
lowQuality	lqEvent
highQuality	hqEvent

Table 3.9: View *Condition*$_1$

Policy	Name	Type
lowQuality	lqCondition	lt
highQuality	hqCondition	gt

Table 3.10: View *ConditionNesting*$_1$

OuterCondition	InnerCondition

Table 3.11: View *Action*$_1$

Policy	Name	No
lowQuality	lqAction	1
highQuality	hqAction	1

3.5 Domain-Specific Policy Modeling

So far, domain and policies have been modeled independently of each other. The domain has been specified as the domain model, and policies have been specified as the policy model. Now, both models must be combined in order to use domain-specific information within the policies. For this purpose, a third model enables policies to refer to the domain in their event, condition, and action parts.

We cover information about how domain-specific information is used within the policies with the linking model, as illustrated in figure 3.23. The linking model specifies how the domain and the policy models are linked to each other. For this purpose, it allows to create links from the objects in the policy model to the objects in the domain model at all layers. Policy experts specify the linking model at design time and can change it at runtime whenever the system behavior should be changed. As with domain and policy modeling, the linking model is an instance of the linking metamodel and offers a particular view at any layer. Views filter the part of the linking model that is visible at the respective layer. Two layers i and j are shown exemplarily in figure 3.23.

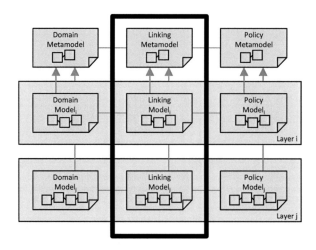

Figure 3.23: Linking policies to the domain

In the following sections, we describe how we use the linking metamodel and the linking model to specify links from the policy model to the domain model. A relational algebra formally defines the linking metamodel, i.e. the abstract syntax of the linking model.

3.5.1 Linking Metamodel

The linking metamodel provides means to create links from the policy model to the domain model, as shown in figure 3.24. Entities of the policy metamodel are highlighted in orange and entities of the domain metamodel in green. The linking metamodel represents the abstract syntax of the links, i.e. it defines the structure of the linking model.

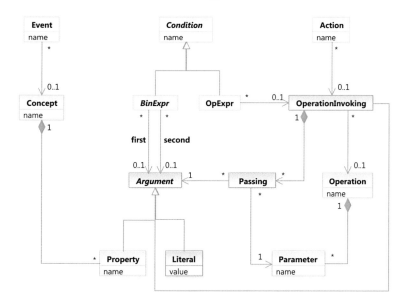

Figure 3.24: Linking metamodel

The linking metamodel defines links from the policy metamodel to the domain metamodel as follows. An *event* can correspond to a *concept*. The *properties* of the concept are used to pass information to the respective policies. The same concept can be used for multiple events. A *condition* in the form of *binary expression* can compare two *arguments* to each other. The simplest form of an argument is a *literal*, i.e. a textual, numeric, boolean, or undefined value. A property can also be used as argument in a binary expression if it is visible to the respective policy, i.e. if the concept of the property is used as event of the policy that uses the binary expression. This allows to use contextual information of the domain within the policy. Another form of an argument is an *operation invoking*. In this case, an *operation* is invoked and the return value of the invoked operation is used as argument. When invoking an operation, arguments can be *passed* to the *parameters* of the operation. In the same way, an *operational expression* can invoke an operation and use the boolean return value of the invoked operation as value of the expression. In order to enforce the desired system behavior, a policy can invoke an operation as its *action*.

Operation invokings are unambiguously identified by their name and their name is used to denote them. Therefore, the name of an operation invoking is unique.

3.5.2 Linking Model

The linking model specifies links from the policy model to the domain model in order to describe which domain-specific information is used in which policies and in which way. These links can only be specified between objects that are modeled at the same abstraction layer, i.e. a policy can only use domain-specific concepts that are visible at the abstraction layers of that policy. As the linking model is an instance of the linking metamodel, various restrictions must hold on its structure.

In the following, we formally define the abstract syntax of the linking model with a relational algebra. Figure 3.25 shows an overview of the respective relations as an ER diagram.

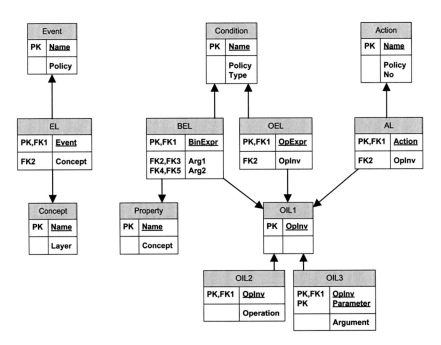

Figure 3.25: ER diagram of the linking model

Definition 15 (*Linking model*)

Let EL be a relation of event links, BEL be a relation of binary expression links, OEL a relation of operational expression links, AL a relation of action links, $OIL1, OIL2, OIL3$ relations of operational expression links.

We define the linking model

$$LM := (EL, BEL, OEL, AL, OIL1, OIL2, OIL3) \tag{3.114}$$

as a septuple of relations that specifies which links from the policy model to the domain model are modeled. A linking model is valid if the model is well-formed according to (3.133), (3.142), (3.155), (3.162), and (3.116)–(3.119), and if the semantic properties hold for it according to (3.134)–(3.136), (3.143)–(3.147), (3.156)–(3.159), (3.163)–(3.166), and (3.120)–(3.123) hold for the enclosed relations.

Let furthermore EL_{la} be a view of EL, BEL_{la} be a view of BEL, OEL_{la} a view of OEL, AL_{la} a view of AL, $OIL1_{la}, OIL2_{la}, OIL3_{la}$ a view of $OIL1, OIL2, OIL3$ at layer la.

We define the view of the linking model LM at layer la

$$LM_{la} := (EL_{la}, BEL_{la}, OEL_{la}, AL_{la}, OIL1_{la}, OIL2_{la}, OIL3_{la}) \tag{3.115}$$

as a septuple of views that specifies which links from the policy model to the domain model are modeled at a particular layer. A view of the linking model is valid if the view is well-formed according to (3.133), (3.142), (3.155), (3.162), and (3.116)–(3.119), and if the semantic properties hold for it according to (3.134)–(3.136), (3.143)–(3.147), (3.156)–(3.159), (3.163)–(3.166), and (3.120)–(3.123) hold for the enclosed views.

Operation Invoking Links

We now formally define operation invoking links with a relational algebra. This includes the entities *OperationInvoking, Operation, Passing, Parameter,* and *Argument* in the linking metamodel and the associations between them (cf. figure 3.24). We perform the formalization with the relations $OIL1$, $OIL2$, and $OIL3$, as illustrated in figure 3.26.

An operation invoking link specifies the invoking of an operation, i.e. which operation is invoked and which arguments are passed to the parameters of the operation. An operation invoking is identified by a name (3.116). The name of an operation invoking represents the PK of the relation. Furthermore, an operation invoking can be linked to one operation (3.117,3.121) that was modeled before (3.120). The respective operation is referenced as FK with its name. An operation invoking can pass arguments to the parameters of the operation (3.118,3.122,3.123). Here, the same parameter can even be passed different parameters at a time. Depending on the underlying system, this may represent set of arguments or a choice of different possible arguments. The respective parameters are referenced as FKs with their names. An argument can be a literal, a property, or the return value of another operation invoking (3.119).

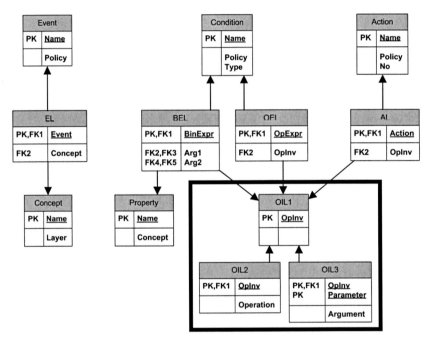

Figure 3.26: Operation invoking links in the ER diagram

Definition 16 (*Operation invoking links*)

Let *Lit* be the set of literals. Let Id_{La} be the set of layer identifiers, Id_{OI} the set of operation invoking identifiers. Let *Pr* the set of modeled properties, *Op* the set of modeled operations, *Pa* the set of modeled parameters. Let $la \in Id_{La}$, $pr \in Pr$, $op, op_1, op_2 \in Op$, $pa \in Pa$, $oi \in Id_{OI}$, $arg, arg_1, arg_2 \in Arg$.

The relations

$$OIL1(OpInv) \subseteq Id_{OI} \tag{3.116}$$

$$OIL2(OpInv, Operation) \subseteq Id_{OI} \times Id_{Op} \tag{3.117}$$

$$OIL3(OpInv, Parameter, Argument) \subseteq Id_{OI} \times Id_{Pa} \times Arg \tag{3.118}$$

$$\text{with } Arg := Lit \cup Id_{Pr} \cup Id_{OI} \tag{3.119}$$

specify which operations are invoked and which arguments are passed to which parameters. Therefore, the following must hold:

$$(oi, op) \in OIL2 \Rightarrow oi \in OIL1 \wedge op \in Op \tag{3.120}$$

$$(oi, op_1) \in OIL2 \wedge (oi, op_2) \in OIL2 \Rightarrow op_1 = op_2 \tag{3.121}$$

$$(oi, pa, arg) \in OIL3 \Rightarrow oi \in OIL1 \wedge pa \in Pa$$
$$\wedge (arg \in Id_{Pr} \Rightarrow arg \in Pr) \wedge (arg \in Id_{OI} \Rightarrow arg \in OIL1) \tag{3.122}$$

$$(oi, pa, arg) \in OIL3 \Rightarrow \exists op.(oi, op) \in OIL2$$
$$\wedge pa \in parametersOfOperation(op) \tag{3.123}$$

The function

$$operationOfOpInv : Id_{OI} \rightarrow Id_{Op}$$

$$oi \mapsto \begin{cases} op, & (oi, op) \in OIL2 \\ undef, & \text{otherwise} \end{cases} \tag{3.124}$$

determines the operation invoked by a particular operation invoking. Due to (3.121), the result is unique.

The function

$$operationsOfOpInv : Id_{OI} \rightarrow \mathcal{P}(Id_{Op})$$

$$oi \mapsto \{op|(oi, op) \in OIL2\}$$
$$\cup \bigcup_{oi_2 \in nestedOpInvs} operationsOfOpInv(oi_2) \tag{3.125}$$

$$\text{with } nestedOpInvs = \{arg|\exists pa.(oi, pa, arg) \in OIL3\} \cap Id_{OI}$$

determines recursively the operations invoked by a particular operation invoking.

The function

$$parametersOfOpInv : Id_{OI} \rightarrow \mathcal{P}(Id_{Pa})$$

$$oi \mapsto \{pa|\exists arg.(oi, pa, arg) \in OIL3\}$$
$$\cup \bigcup_{oi_2 \in nestedOpInvs} parametersOfOpInv(oi_2) \tag{3.126}$$

$$\text{with } nestedOpInvs = \{arg|\exists pa.(oi, pa, arg) \in OIL3\} \cap Id_{OI}$$

determines recursively the properties that are passed in a particular operation invoking. The intersection with the set of operation invoking identifiers eliminates undesired types of arguments.

The function

$$propertiesOfOpInv : Id_{OI} \rightarrow \mathcal{P}(Id_{Pr})$$
$$oi \mapsto \{arg | \exists pa.(oi, pa, arg) \in OIL3\} \cap Id_{Pr}$$
$$\cup \bigcup_{oi_2 \in nestedOpInvs} propertiesOfOpInv(oi_2) \quad (3.127)$$

$$\text{with } nestedOpInvs = \{arg | \exists pa.(oi, pa, arg) \in OIL3\} \cap Id_{OI}$$

determines recursively the properties that are passed in a particular operation invoking. The intersections with the sets of properties and operation invoking identifiers eliminates undesired types of arguments.

The function

$$layersOfOpInv : Id_{OI} \rightarrow \mathcal{P}(Id_{La})$$
$$oi \mapsto \bigcap_{op \in operationsOfOpInv(oi)} layersOfOperation(op)$$
$$\cap \bigcap_{pa \in parametersOfOpInv(oi)} layersOfParameter(pa) \quad (3.128)$$
$$\cap \bigcap_{pr \in propertiesOfOpInv(oi)} layersOfProperty(pr)$$

determines the layers where a particular operation invoking is available. An operation invoking is available only at those layers where all linked operations and passed properties are available in common.

The function

$$opInvsOfLayer : Id_{La} \rightarrow \mathcal{P}(Id_{OI})$$
$$la \mapsto \{oi | oi \in OIL1 \wedge la \in layersOfOpInv(oi)\} \quad (3.129)$$

determines the available operation invokings at a particular layer.

The views

$$OIL1_{la} := \{oi | oi \in OIL1$$
$$\wedge oi \in opInvsOfLayer(la)\} \subseteq OIL1 \quad (3.130)$$

$$OIL2_{la} := \{(oi, op) | (oi, op) \in OIL2$$
$$\wedge oi \in opInvsOfLayer(la)\} \subseteq OIL2 \quad (3.131)$$

$$OIL3_{la} := \{(oi, pa, arg) | (oi, pa, arg) \in OIL3$$
$$\wedge oi \in opInvsOfLayer(la)\} \subseteq OIL3 \quad (3.132)$$

show the available operation invokings at a particular layer.

Event Links

We now formally define event links with a relational algebra. This includes the entities *Event* and *Concept* in the linking metamodel and the association between them (cf. figure 3.24). We perform the formalization with the relation *EL*, as illustrated in figure 3.27.

An event link specifies which concept is used as event. For this purpose, an event is linked to one concept (3.133,3.135), and both must have been modeled before (3.134). The respective event and concept are referenced as FKs with their names. The event represents the PK of the relation. A concept can be linked from multiple events (3.133).

An important constraint comes from the fact that events are used at particular layers, but the linked concepts are available at other layers. For a valid linking model, an event can only be linked to a concept if the concept is available at all layers where the event is used (3.136).

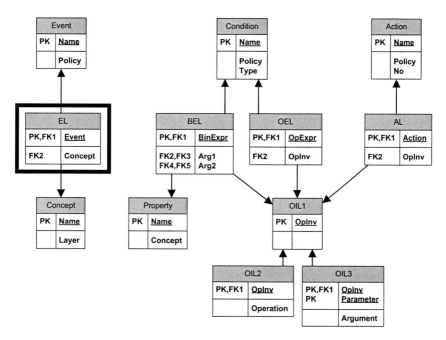

Figure 3.27: Event links in the ER diagram

Definition 17 (*Event links*)

Let Id_{La} be the set of layer identifiers, Id_{Co} the set of concept identifiers, Id_{Po} the set of policy identifiers, Id_{Ev} the set of event identifiers. Let

Co be the set of modeled concepts, Ev the set of modeled events. Let $la \in Id_{La}, co, co_1, co_2 \in Id_{Co}, po \in Id_{Po}, ev \in Id_{Ev}$.

The relation

$$EL(Event, Concept) \subseteq Id_{Ev} \times Id_{Co} \tag{3.133}$$

specifies which concepts represent which events. Therefore, the following must hold:

$$(ev, co) \in EL \Rightarrow ev \in Ev \wedge co \in Co \tag{3.134}$$

$$(ev, co_1) \in EL \wedge (ev, co_2) \in EL \Rightarrow co_1 = co_2 \tag{3.135}$$

$$(ev, co) \in EL \Rightarrow layersOfEvent(ev) \subseteq layersOfConcept(co) \tag{3.136}$$

The function

$$conceptOfEvent : Id_{Ev} \rightarrow Id_{Co}$$

$$ev \mapsto \begin{cases} co, & (ev, co) \in EL \\ undef, & \text{otherwise} \end{cases} \tag{3.137}$$

determines the concept that corresponds to a particular event. Due to (3.135), the result is unique.

The function

$$visibleConcepts : Id_{Po} \rightarrow \mathcal{P}(Id_{Co})$$

$$po \mapsto \bigcup_{ev \in eventsOfPolicy(po)} accessibleConcepts(conceptOfEvent(ev)) \tag{3.138}$$

determines the concepts that are visible to a particular policy. The visible concepts are the concepts that are accessible form the concepts that represent the events of the policy.

The function

$$visibleProperties : Id_{Po} \rightarrow \mathcal{P}(Id_{Pr})$$

$$po \mapsto \bigcup_{co \in visibleConcepts(po)} propertiesOfConcept(co) \tag{3.139}$$

determines the properties that are visible to a particular policy. The visible properties are the properties of the visible concepts.

The view

$$EL_{la} := \{(ev, co) | (ev, co) \in EL \\ \wedge ev \in eventsOfLayer(la)\} \subseteq EL \tag{3.140}$$

shows the available event links at a particular layer.

Binary Expression Links

We now formally define binary expression links with a relational algebra. This includes the entities *BinExpr* and *Argument* in the linking metamodel and the associations between them (cf. figure 3.24). We perform the formalization with the relation *BEL*, as illustrated in figure 3.28.

A binary expression link specifies which arguments are used for the comparison. For this purpose, a binary expression is linked to two arguments (3.142,3.144). An argument can be a literal, a property, or an operation invoking (3.119), and the latter two must have been modeled before (3.143). The respective binary expression and the properties and operation invokings within the arguments are referenced as FKs with their names. The binary expression represents the PK of the relation. An argument can be linked from multiple binary expressions (3.142).

An important constraint for a valid linking model comes from the separation of information into different models and layers. A policy can only use information from the domain that is visible to this policy. This restricts the usage of properties and operations in policy conditions and especially in binary expression links, as these create links from the policy condition to the domain. Firstly, a property can only be used in a condition if it is visible to all policies that use this condition (3.145,3.146). A property is visible to a policy if it is associated to the concept representing a policy event or to a concept that is accessible from this concept via relationships. Secondly, properties and operation invokings can only be used in a condition if they are available at all layers where this condition is used (3.147).

Definition 18 (*Binary expression links*)

Let Id_{Po} be the set of policy identifiers, Id_{Cd} the set of condition identifiers, $Type_{Cd}$ the set of condition types. Let *Condition* be a set of conditions. Let $po \in Id_{Po}$, $cd \in Id_{Cd}$, $ty \in Type_{Cd}$.

The set

$$Be := \{cd | \exists po, ty.(po, cd, ty) \in Condition$$
$$\wedge\ ty \in \{eq, gt, ge, lt, le\}\} \subseteq Id_{Cd} \tag{3.141}$$

contains all modeled binary expressions.

Let furthermore Id_{La} be the set of layer identifiers, Id_{Pr} the set of property identifiers, Id_{OI} the set of operation invoking identifiers. Let $la \in Id_{La}$, $pr \in Id_{Pr}$, $be \in Id_{Cd}$, $oi \in Id_{OI}$, $arg_1, arg_2, arg_3, arg_4 \in Arg$.

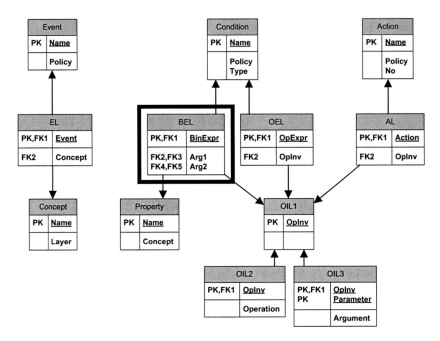

Figure 3.28: Binary expression links in the ER diagram

The relation

$$BEL(BinExpr, Arg1, Arg2) \subseteq Id_{Cd} \times Arg \times Arg \qquad (3.142)$$

specifies which binary expression compares which arguments. Therefore, the following must hold:

$$(be, arg_1, arg_2) \in BEL \Rightarrow be \in Be$$
$$\wedge (arg_1 \in Id_{Pr} \Rightarrow arg_1 \in Pr) \wedge (arg_2 \in Id_{Pr} \Rightarrow arg_2 \in Pr) \qquad (3.143)$$
$$\wedge (arg_1 \in Id_{OI} \Rightarrow arg_1 \in OIL1) \wedge (arg_2 \in Id_{OI} \Rightarrow arg_2 \in OIL1)$$

$$(be, arg_1, arg_2) \in BEL \wedge (be, arg_3, arg_4) \in BEL$$
$$\Rightarrow arg_1 = arg_3 \wedge arg_2 = arg_4 \qquad (3.144)$$

$$(be, arg_1, arg_2) \in BEL$$
$$\Rightarrow (arg_1 \in Id_{Pr} \Rightarrow \forall po \in policiesOfCondition(be).$$
$$arg_1 \in visibleProperties(po)) \qquad (3.145)$$
$$\wedge (arg_2 \in Id_{Pr} \Rightarrow \forall po \in policiesOfCondition(be).$$
$$arg_2 \in visibleProperties(po))$$

$(be, arg_1, arg_2) \in BEL$
$\Rightarrow (arg_1 \in Id_{OI} \Rightarrow \forall po \in policiesOfCondition(be)$
$\forall pr \in propertiesOfOpInv(arg_1).$
$pr \in visibleProperties(po))$ (3.146)
$\wedge (arg_2 \in Id_{OI} \Rightarrow \forall po \in policiesOfCondition(be)$
$\forall pr \in propertiesOfOpInv(arg_2).$
$pr \in visibleProperties(po))$

$(be, arg_1, arg_2) \in BEL \Rightarrow layersOfCondition(be)$
$$\subseteq \bigcap_{pr \in propertiesOfBinExpr(be)} layersOfProperty(pr)$$ (3.147)
$$\cap \bigcap_{oi \in opInvsOfBinExpr(be)} layersOfOpInv(oi)$$

The function

$$argument1OfBinExpr : Id_{Cd} \rightarrow Arg$$

$$be \mapsto \begin{cases} arg_1, & \exists arg_2.(be, arg_1, arg_2) \in BEL \\ undef, & otherwise \end{cases}$$ (3.148)

determines the first argument of a particular binary expression. Due to (3.144), the result is unique.

The function

$$argument2OfBinExpr : Id_{Cd} \rightarrow Arg$$

$$be \mapsto \begin{cases} arg_2, & \exists arg_2.(be, arg_1, arg_2) \in BEL \\ undef, & otherwise \end{cases}$$ (3.149)

determines the second argument of a particular binary expression. Due to (3.144), the result is unique.

The function

$$literalsOfBinExpr : Id_{Cd} \rightarrow \mathcal{P}(Lit)$$
$$be \mapsto \{argument1OfBinExpr(be), argument2OfBinExpr(be)\}$$ (3.150)
$$\cap Lit$$

determines the literals that are used within a particular binary expression. The intersection with the set of literals eliminates undesired types of arguments.

The function

$$propertiesOfBinExpr : Id_{Cd} \rightarrow \mathcal{P}(Id_{Pr})$$
$$be \mapsto \{argument1OfBinExpr(be), argument2OfBinExpr(be)\}$$ (3.151)
$$\cap Id_{Pr}$$

determines the properties that are referred by a particular binary expression. The intersection with the set of property identifiers eliminates undesired types of arguments.

The function

$$opInvsOfBinExpr : Id_{Cd} \rightarrow \mathcal{P}(Id_{OI})$$
$$be \mapsto \{argument1OfBinExpr(be), argument2OfBinExpr(be)\} \quad (3.152)$$
$$\cap Id_{OI}$$

determines the operation invokings referred directly by a particular binary expression and includes the operation invokings that are nested in the initial operation invoking. The intersection with the set of operation invoking identifiers eliminates undesired types of arguments.

The view

$$BEL_{la} := \{(be, arg_1, arg_2) | (be, arg_1, arg_2) \in BEL$$
$$\wedge be \in conditionsOfLayer(la)\} \subseteq BEL \quad (3.153)$$

shows the available binary expression links at a particular layer.

Operational Expression Links

We now formally define operational expression links with a relational algebra. This includes the entities *OpExpr* and *OperationInvoking* in the linking metamodel and the association between them (cf. figure 3.24). We perform the formalization with the relation *OEL*, as illustrated in figure 3.29.

An operational expression link specifies which operation is invoked for the evaluation of a condition. For this purpose, an operational expression is linked to one operation invoking (3.155,3.157), and both must have been modeled before (3.156). The respective condition and operation invoking are referenced as FKs with its names. The operational expression represents the PK of the relation. An operation invoking can be linked from multiple operational expressions (3.155).

Operational expression links create links from policy conditions to the domain. For a valid linking model, the usage of properties and operations in operational expression links is restricted, similarly to the usage of arguments in binary expression links. Firstly, a property can only be used in a condition if it is visible to all policies that use this condition (3.158). A property is visible to a policy if it is associated to the concept representing the policy event or to a concept that is accessible via relationships. Secondly, properties and operation invokings can only be used in a condition if they are available at all layers where this condition is used (3.159).

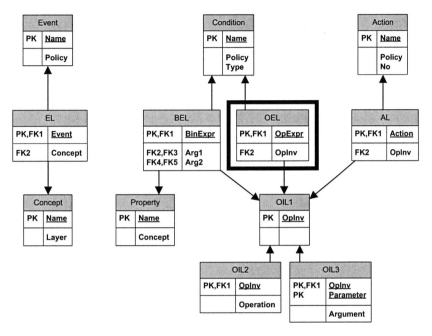

Figure 3.29: Operational expression links in the ER diagram

Definition 19 (*Operational expression links*)

Let Id_{Po} be the set of policy identifiers, Id_{Cd} the set of condition identifiers, $Type_{Cd}$ the set of condition types. Let *Condition* be a set of conditions. Let $po \in Id_{Po}$, $cd \in Id_{Cd}$, $ty \in Type_{Cd}$.

The set

$$Oe := \{cd | \exists po, ty.(po, cd, ty) \in Condition \land ty \in \{op\}\} \subseteq Id_{Cd} \qquad (3.154)$$

contains all modeled operational expressions.

Let furthermore Id_{La} be the set of layer identifiers, Id_{Pr} the set of property identifiers, Id_{Cd} the set of condition identifiers, Id_{OI} the set of operation invoking identifiers. Let $OIL1$ be a first relation of operation invokings. Let $la \in Id_{La}$, $pr \in Id_{Pr}$, $oe \in Id_{Cd}$, $oi, oi_1, oi_2 \in Id_{OI}$.

The relation

$$OEL(OpExpr, OpInv) \subseteq Id_{Cd} \times Id_{OI} \qquad (3.155)$$

specifies which operational expression invokes which operation and which arguments it passes to the parameters of the operation. Therefore, the following must hold:

$$(oe, oi) \in OEL \Rightarrow oe \in Oe \wedge oi \in OIL1 \qquad (3.156)$$

$$(oe, oi_1) \in OEL \wedge (oe, oi_2) \in OEL \Rightarrow oi_1 = oi_2 \qquad (3.157)$$

$$(oe, oi) \in OEL \Rightarrow \forall po \in policiesOfCondition(oe) \qquad (3.158)$$
$$\forall pr \in propertiesOfOpInv(oi).pr \in visibleProperties(po)$$

$$(oe, oi) \in OEL \Rightarrow layersOfCondition(oe) \subseteq layersOfOpInv(oi) \quad (3.159)$$

The function

$$opInvOfOpExpr : Id_{Cd} \rightarrow Id_{OI}$$

$$oe \mapsto \begin{cases} oi, & (oe, oi) \in OEL \\ undef, & \text{otherwise} \end{cases} \qquad (3.160)$$

determines the operation invoking that corresponds to a particular operational expression. Due to (3.157), the result is unique.

The view

$$OEL_{la} := \{(oe, oi) | (oe, oi) \in OEL$$
$$\wedge oe \in conditionsOfLayer(la)\} \subseteq OEL \qquad (3.161)$$

shows the available operational expression links at a particular layer.

Action Links

We now formally define action links with a relational algebra. This includes the entities *Action* and *OperationInvoking* in the linking metamodel and the association between them (cf. figure 3.24). We perform the formalization with the relation *AL*, as illustrated in figure 3.30.

An action link specifies which operation is invoked as action. For this purpose, an action is linked to one operation invoking (3.162,3.164), and both must have been modeled before (3.163). The respective action and operation invoking are referenced as FKs with their names. The action represents the PK of the relation. An operation invoking can be linked from multiple actions (3.162).

Action links create links from policy actions to the domain. For a valid linking model, the usage of properties and operations in operational expression links is restricted, similarly to the usage of arguments in binary and operational expression links. Firstly, a property can only be used in an action if it is visible to all policies that use this action (3.165). A property is visible to a policy if it is associated to the concept used as policy event or to a concept that is accessible via relationships. Secondly, an operation invoking can only be used in an action if it is available at all layers where this action is used (3.166).

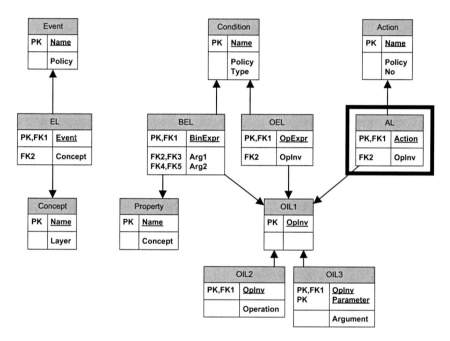

Figure 3.30: Action links in the ER diagram

Definition 20 (*Action links*)

Let Id_{La} be the set of layer identifiers, Id_{Pr} the set of property identifiers, Id_{Po} the set of policy identifiers, Id_{Ac} the set of action identifiers, Id_{OI} the set of operation invoking identifiers. Let Ac be the set of modeled actions. Let $OIL1$ be a first relation of operation invokings. Let $la \in Id_{La}$, $pr \in Id_{Pr}$, $po \in Id_{Po}$, $ac \in Id_{Ac}$, $oi, oi_1, oi_2 \in Id_{OI}$.

The relation

$$AL(Action, OpInv) \subseteq Ac \times Id_{OI} \tag{3.162}$$

specifies which action invokes which operation and which arguments it passes to the parameters of the operation.

Therefore, the following must hold:

$$(ac, oi) \in AL \Rightarrow ac \in Ac \wedge oi \in OIL1 \tag{3.163}$$

$$(ac, oi_1) \in AL \wedge (ac, oi_2) \in AL \Rightarrow oi_1 = oi_2 \tag{3.164}$$

$$(ac, oi) \in AL \Rightarrow \forall po \in policiesOfAction(ac)$$
$$\forall pr \in propertiesOfOpInv(oi).pr \in visibleProperties(po) \tag{3.165}$$

$$(al, oi) \in AL$$
$$\Rightarrow layersOfAction(ac) \subseteq layersOfOpInv(oi) \tag{3.166}$$

The function

$$opInvOfAction : Id_{Ac} \rightarrow Id_{OI}$$
$$ac \mapsto \begin{cases} oi, & (ac, oi) \in AL \\ undef, & \text{otherwise} \end{cases} \tag{3.167}$$

determines the operation invoking that corresponds to a particular action. Due to (3.164), the result is unique.

The view

$$AL_{la} := \{(ac, oi) | (ac, oi) \in AL$$
$$\wedge ac \in actionsOfLayer(la)\} \subseteq AL \tag{3.168}$$

shows the available action links at a particular layer.

3.5.3 Example

In the following, we present a linking model that specifies the links from the example policy model provided in section 3.4.3 to the example domain model provided in section 3.3.3. We model the links at the first abstraction layer only. The links at the second layer are generated by an automated refinement of the first layer, as shown in section 4.2.4. Figure 3.31 shows an overview of the linking model with a concrete syntax based on UML. Tables 3.12 to 3.18 provide the formal representation of the linking model.

We now represent the events *lqEvent* and *hqEvent* as the concept *signalQuality*. In order to analyze the variation of signal quality, the conditions *lqCondition* and *hqCondition* compare the property *sqNewValue* to respective literals *50* and *80*. Finally, the actions *lqAction* and *hqAction* invoke the operations *increasePower* and *decreasePower*.

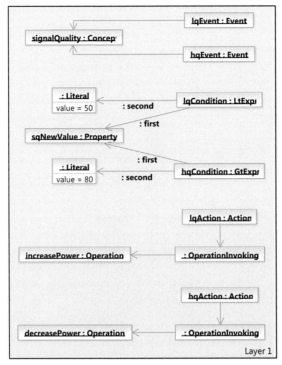

Figure 3.31: Example linking model (layer 1)

Table 3.12: View EL_1

Event	Concept
lqEvent	signalQuality
hqEvent	signalQuality

Table 3.13: View AL_1

Action	OpInv
lqAction	lqActionOpInv
hqAction	hqActionOpInv

Table 3.14: View BEL_1

BinExpr	Arg1	Arg2
lqCondition	sqNewValue	50
hqCondition	sqNewValue	80

Table 3.15: View OEL_1

OpExpr	OpInv

Table 3.16: View $OIL1_1$

OpInv
lqActionOpInv
hqActionOpInv

Table 3.17: View $OIL2_1$

OpInv	Operation
lqActionOpInv	increasePower
hqActionOpInv	decreasePower

Table 3.18: View $OIL3_1$

OpInv	Parameter	Argument

3.6 Related Work

In this section, we summarize approaches to the specification of policies that are comparable to our approach. We first describe each approach and finally outline their commonalities and differences in a comparison table. Moreover, we compare them to the features of our approach.

General Policy Modeling Language

The authors of [105] present a model-driven approach to the design of policies and their integration into the software development process. The intention is to represent policies at design time with a focus on logical concepts and to specify system behavior prior to its implementation. Another objective is the sharing and interchange of policies between different engines. For this purpose, the General Policy Modeling Language (GPML) abstracts common policy concepts from several policy languages and allows to deploy policies into an interchange format.

The approach is based on MDE concepts and consists of several components. A metamodel based on the Meta Object Facility (MOF) by the Object Management Group (OMG) [162] defines the abstract syntax of GPML. A UML Profile is used to define a concrete graphical syntax. In addition, a concrete textual syntax based on the Extensible Markup Language (XML) [163] is used to interchange GPML policies with other systems. A set of model transformations enables the mapping of GPML policies into a rule interchange format. These transformations are bidirectional, and this allows to reverse engineer existing policies into GPML models. The REWERSE I1 Rule Markup Language (R2ML) [164] is used as interchange format as it also has a concrete graphical syntax. Furthermore, it provides a number of transformations into different policy languages such as Ponder [18], KAoS [160], and Rei [161]. The GPML metamodel is shown in figure 3.32.

The GPML metamodel is an extension of the R2ML metamodel and grounded on the theoretical foundation of deontic logic [165]. The green classes of the metamodel are used for reasons of compatibility with R2ML. A policy rule is generally defined as an R2ML *DerivationRule*. The metamodel provides support for the four different policy types *permission*, *prohibition*, *obligation*, and *dispensation*. Any policy has a *Context* to which it is applied. The association *triggeredBy* represents the event upon whose occurrence the policy is fired. The condition part is composed of R2ML *Logical-Formulas*. For each policy, the *ObjectDescriptionAtom* holds the description in the deontic logic. The associations *hasAction* and *hasEffect* relate a policy to its action and to the effects of applying this policy. The actor is specified by the association *performedBy*. The association *obliges* is specific to obligation

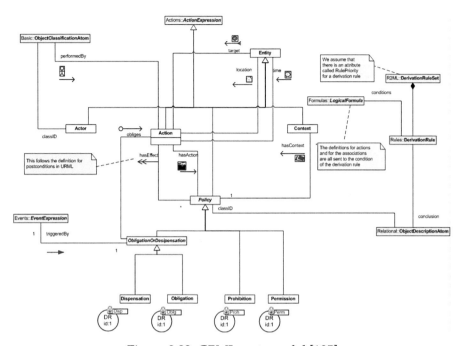

Figure 3.32: GPML metamodel [105]

policies and represents the obligatory task an actor ought to perform upon policy execution. Finally, the metamodel shows symbols to graphically represent GPML constructs in the UML profile.

Common Information Model

The Distributed Management Task Force (DMTF) is a large industry organization leading the development, adoption, and promotion of interoperable management standards and initiatives. They focus on management interoperability among multi-vendor systems, tools, and solutions within an enterprise and aim at reducing management complexity. To cover various aspects of a managed environment, the DMTF defined a set of object-oriented information models. The Common Information Model (CIM) [106] represents a conceptual framework for describing a system architecture and the system entities to be managed. An extension to the CIM to describe policies and define policy control is provided by the CIM Policy Model [107]. This model was developed after the CIM, and its development was heavily influenced by the Internet Engineering Task Force (IETF) Policy Core Information Model (PCIM) [166]. However, work on the PCIM was not continued by the IETF. The DMTF models have reached a rather stable state. However, updates are released from time to time including minor changes.

The CIM Policy Model by the DMTF addresses the management of complex multi-vendor environments with a huge number of heterogeneous devices. It represents a metamodel that enables administrators to specify policies in a device-independent way by abstracting from hardware characteristics and representing heterogeneous devices as managed entities. Policies are specified in a declarative way omitting technical details. The model focuses on network management and addresses a variety of scenarios such as Quality of Service (QoS) and IP Security (IPSec). An excerpt of the CIM Policy Model is shown in figure 3.33.

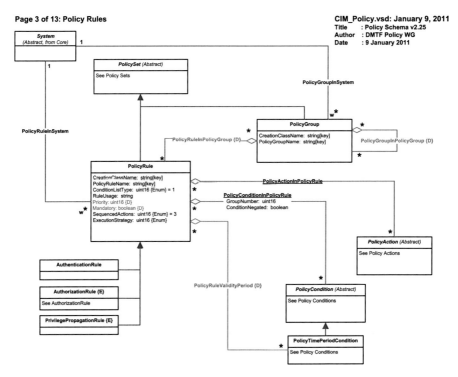

Figure 3.33: Excerpt of the CIM Policy Model [108]

In the CIM Policy Model, *PolicySet* is an abstract concept that cannot be instantiated, but encapsulates common attributes, methods, and associations of the two containers *PolicyRule* and *PolicyGroup*. *PolicyRule* is an aggregation of *PolicyActions* and *PolicyConditions*. A *PolicyGroup* aggregates *PolicyRules* and other *PolicyGroups* and thus allows for introducing a hierarchy of *PolicyRules*. In this model, no events are specified as the focus is on authorization policies that are permanently valid. The CIM Policy Model supports both single and aggregated policies and offers boolean operations in order to combine several conditions to more complex constructs.

Directory Enabled Networks - Next Generation

The TM Forum is a large trade association whose aim is to improve the way telecommunication, media, and information services are created, delivered, assured, and charged. They also facilitate marketing and networking activities for their members. The TM Forum developed another type of information model called the Directory Enabled Networks - next generation (DEN-ng) [109–111]. The development of this model was influenced by their New Generation Operational Systems and Software (NGOSS) architecture [167] and the Shared Information and Data Model (SID) [168].

The concept of DEN-ng is similar to the DMTF approach. However, DEN-ng models address the modeling of various domains and policies within these domains in a uniform and consistent way. They also address the abstraction of network management, even though this is not the primary focus. DEN-ng was developed to integrate information models and management architectures in order to describe how different management information in a system is related to each other. It represents an information model that offers means to represent various entities and policies within a managed environment. The DEN-ng metamodel is based on the policy continuum and considers different levels of abstraction. In order to align the terminology of the different abstraction levels, the SID is used as common vocabulary shared by all levels. Policies are not an extension, but directly integrated into the models. Figure 3.34 shows an excerpt of DEN-ng that is relevant for policies.

The concepts are similar to the CIM Policy Model. *PolicySet* is an abstract concept that cannot be instantiated, but encapsulates commonalities of *PolicyRule* and *PolicyGroup*. The name *PolicySet* also ensures backward compatibility with the IETF models. *PolicyGroup* is only used as a container that collects *PolicyRules* and other *PolicyGroups*. Its aggregation *ContainedPolicySets* allows for introducing a hierarchy of policy rules. The cardinalities of this aggregation enable a policy group to contain an arbitrary number of other policy groups and rules, and they ensure that each of these is contained in not more than one group. Loops where a policy group contains itself are not allowed. *PolicyRule* represents ECA policies in DEN-ng. A *PolicyRule* contains at least one *PolicyEventSet*, one *PolicyCondition*, and one *PolicyAction*, whereas each of these can be assigned to multiple *PolicyRules* or can exist without being bound to a *PolicyRule* at all. In order to handle an arbitrary number of events at a time, *PolicyEvents* are not assigned to policy rules directly, but via *PolicyEventSets*. *PolicyEventSets* control the execution of one or more *PolicyGroups* by triggering the policies enclosed in these groups. The cardinalities of the association *ControlsExecutionOf* also make sure that the whole group is controlled by at most one event set, and this prohibits the grouping of policy rules that are controlled by different event sets.

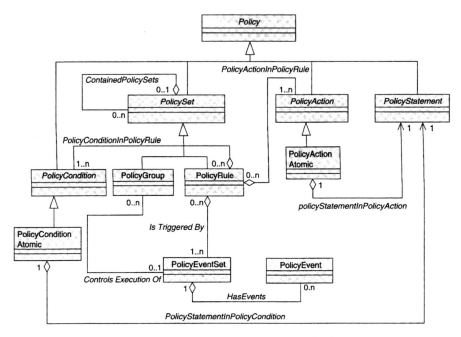

Figure 3.34: Excerpt of DEN-ng [19]

Comparison

The presented approaches to the modeling of policies differ in some aspects from our approach. We outline the differences in table 3.19. A + in the table indicates that a feature is available in an approach, a *o* indicates a limited feature, and a - indicates that a feature is not available at all.

At first, there are differences between the presented approaches with regard to the supported policy types. GPML supports ECA policies with the obligation policy type. DEN-ng has integral support for ECA policies, although other types are supported as well. We explicitly support ECA policies in our approach. In contrast, the CIM Policy Model focuses on authorization and does not support policies that are triggered by events. At least, extensions to this model allow to represent events within conditions and to distinguish triggering events from other conditions.

Policies in GPML can be specified in a customized domain due to the ability of the underlying R2ML to define a particular vocabulary. Similarly, DEN-ng enables the specification of a domain and resources within that domain on which policies operate. Our domain metamodel allows to specify policies with respect to a customized domain as well. In contrast, the CIM is a domain-specific model with a focus on network management. Thus, policies in the CIM Policy Model cannot be applied to other domains.

Table 3.19: Comparison of policy specification approaches

	GPML [105]	CIM [106–108]	DEN-ng [109–111]	Model-Based Policy Specification
ECA policies	+	o	+	+
Customizable domain	+	-	+	+
Abstraction levels	-	-	+	+
Abstract syntax	+	+	+	+
Concrete syntax	+	+	+	o
Language-independent	+	+	+	+
Persistent storage	-	-	-	+
Formal semantics	-	-	-	+

Policies are considered at different levels of abstraction by DEN-ng, and DEN-ng even supports a flexible number of levels. The same applies to our approach. In contract, GPML policies must initially be modeled at a low level of abstraction in order to enable their transformation into other languages. The CIM Policy Model does not allow for different abstraction levels either. It does not consider different levels as it only addresses the specification of policies, and it does not deal with deploying policies to a certain platform nor with generating a device-specific representation of policies.

The abstract syntax of policies is defined by a metamodel in all approaches. There a minor differences with regard to their concrete syntax. GPML policies can be modeled in XML, but also graphically in UML as a UML profile is provided. The CIM also provides a UML profile for the graphical representation of policies. The DEN-ng metamodel is defined in UML, and policies are represented as class diagrams in UML. In contrast, we provide a relational algebra as a basis for possible concrete syntaxes and are developing a concrete textual syntax.

All approaches are independent of any particular policy language or engine and enable the modeling of policies in a language-independent way. Their metamodels allow the developer to describe a system and the policies to manage this system in an implementation-independent way. The approaches specify declarative policies that omit technical details of the underlying system. However, the other approaches do not address a persistent storage of the models nor provide their static semantics in a formal way.

3.7 Conclusion and Outlook

Modern software development is driven by model-based techniques today. Source code is no more simply implemented by hand, but large parts are generated automatically with techniques from MDE. Such automation is achieved with the help of models that are not only used for purposes of documentation, but that represent primary artifacts in the development process. However, model-based solutions are not widely used in Policy-Based Management (PBM), or they are limited and tailored to a particular problem. For this reason, we presented an approach to the specification of policies that is driven by modern paradigms of MDE. We use a flexible combination of different models and different abstraction layers to enable the modeling of domain-specific policies in a generic and powerful way.

Models serve different purposes in our approach. First of all, models raise the level of abstraction as they hide technical details and focus on the relevant aspects. The usage of models to represent policies also eliminates the dependency on a particular policy language and platform as models separate policies from their implementation. This allows to specify policies at an early stage of development already. Our metamodels define the abstract syntax of domain-specific policies. We use a concrete syntax based on UML to visualize the example models in a graphical way as diagrams. However, experience showed that a purely graphical concrete syntax for the models might be confusing as diagrams take a lot of space in more complex scenarios. For this reason, an effective textual syntax is being developed. The formal representation of specific models for purposes of their validation should then automatically be generated from their concrete syntax.

Our approach enables the modeling of domain-specific policies at different levels of abstraction. The usage of models at different layers facilitates the collaboration of experts with a high-level application focus and a low-level technical background. A flexible number of abstraction layers allows these experts to specify policies in their individual terminology. This applies to simple and complex policy-based systems at the same time. Furthermore, we use different models to separate the structural aspects of policies from their domain-specific content. The domain model enables domain experts to specify which concepts are available in a system, the policy model enables policy experts to specify which policies are used to manage the system, and the linking model allows to integrate both models. The separation of different concerns into different models reduces complexity and facilitates the collaboration of domain and policy experts. Both expert groups can initially focus on their background and use their expertise to independently specify the domain and policy models. The linking model then provides a specific interface for both groups to combine their models into the desired domain-specific policies.

We provide a formal definition of the models with a relational algebra. This formalization serves different purposes. Firstly, it defines the abstract syntax of domains and policies. The domain metamodel does not reach the whole expressiveness of an ontology, but provides the relevant aspects needed for the specification of a domain as used in this thesis. The policy metamodel is expressive enough to specify the structure of ECA policies. The linking metamodel provides a flexible means to integrate domain-specific information within policies. Secondly, the formalization defines the static semantics of ECA policies and allows for their validation. It requires that any policy only uses domain-specific information within its condition and action parts that is available via its events. It also requires that any policy only uses information that is available at the layers of the policy. Validation of the models is a prerequisite for their refinement into a lower level of abstraction and their transformation into executable code. The checking of properties and detection of inconsistencies within the set of policies are subjects to future work. Furthermore, the formalization of the models is useful in scenarios where persistent storage should be addressed as their formal representation can directly be mapped to a relational database. Last but not least, the formalization represents a prerequisite for defining the formal semantics of the policy refinement process, which we present in the following chapter.

As our approach is based on metamodels, it is extensible by extending the metamodels. One example is the lacking support for a multiple inheritance of concepts in the current domain metamodel, which would enable the assignment of multiple superconcepts to the same subconcept. Multiple inheritance can be introduced simply by changing the cardinality of the respective association. Of course, this change must also be represented within the formalization of the models. Furthermore, the policy metamodel enables policies according to the ECA paradigm. Other policy types such as goal policies have a different structure and cannot be expressed at the moment. However, additional policy types can be introduced by extending the policy metamodel. Both proposed extensions of the domain and policy metamodels would not affect existing domain or policy models as expressiveness is not restricted in any way, but only increased. For this reason, existing models would still be valid instances of the extended metamodels.

Last but not least, our approach is generic with respect to the domain, to the language, and to the number of abstraction levels. It allows to model domain-specific policies in a language-independent way and at an arbitrary number of layers. Our approach can be applied to a variety of domains, and it is novel in this regard as no approach is known to us that is flexible to such an extent and formally defined at the same time.

4

Model-Driven Policy Refinement

The development of policies is a manual process today that involves different expert groups. Initially, business experts use natural language constructs to specify policies at a high level of abstraction and from a functional point of view. Application experts then add specific functional knowledge to refine these high-level business policies into an application point of view and to make them executable. Afterwards, technical experts implement these policies in a language that is executable by the underlying system. Finally, tests are performed on whether the resulting system behavior meets the objectives of the initial business and intermediate application policies. This process is repeated until the desired behavior is achieved.

The manual refinement of policies is contrary to the paradigms of modern software development. Firstly, collaboration of different expert groups with different background is time-consuming and error-prone. Secondly, the manual implementation of policies is decoupled from their initial requirements and thus imposes additional effort. In this chapter, we present an innovative approach to the automated refinement of policies, as illustrated in figure 4.1. We base our refinement approach on the model-based specification of policies presented in the previous chapter and make use of the separation of concerns into different models and abstraction layers. We automate the refinement process with model transformations that describe mappings between high-level and low-level models and refine the abstract models to more technical ones. The automated refinement allows to instantly reflect changes at the abstract models in their technical representation, and this allows to manage the behavior of the system at a high level of abstraction. Changes at the models are even possible at runtime.

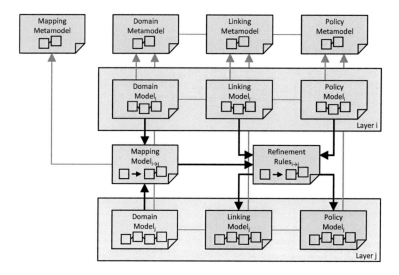

Figure 4.1: Model-driven policy refinement

Instead of refining high-level policies manually, we automate the refinement process by specifying the refinement once within the domain and then applying this knowledge to the respective policies wherever and whenever necessary. For this purpose, domain experts specify in the mapping model how to refine a higher-layer representation of the domain model into a lower-layer one. This is performed by establishing cross-layer mappings between the domain-specific objects from higher to lower layers. Then, refinement rules apply these mappings to the high-level policy and linking models and generate the respective low-level policies in an automated way. This is performed from top to bottom across all abstraction layers, starting with the policies at the highest layer until they are represented at the lowest layer from a technical point of view. This technical policy representation at the lowest layer represents the basis for generating executable code. We use a relational algebra to formally define the refinement rules, i.e. the semantics of the refinement process. Furthermore, the relational algebra allows to prove that the refinement rules preserve the static semantics of the models, i.e. that valid models are refined into valid models again.

In the following sections, we describe the refinement process and the way to automate it. Section 4.1 describes the manual refinement of the domain, and section 4.2 describes how we use this information to generate refined policies in an automated way. Finally, section 4.3 presents related work, and section 4.4 provides a conclusion and an outlook.

4.1 Domain Refinement

As mentioned before, different abstraction layers enable different expert groups to have different views of the domain and to specify domain knowledge in their individual terminology. The view of the domain model at a lower layer usually extends the view at a higher layer. We now present how to successively extend domain knowledge at lower layers in order to refine the functional view of the domain model into a technical view.

We base the refinement of the domain on the formal definition of the domain model, as illustrated in figure 4.2. For this purpose, we cover any relevant information about the domain refinement with the mapping model. The mapping model allows to establish mappings from the view of the domain model at a higher layer to the view of the domain model at the subsequent lower layer. Two successive layers i and j are shown exemplarily in figure 4.2. The mapping model is an instance of the mapping metamodel, which defines the structure of the mapping model.

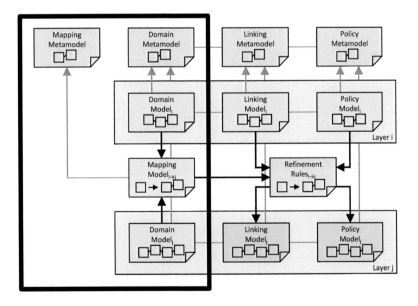

Figure 4.2: Domain refinement

In the following sections, we describe how we use the mapping metamodel and the mapping model to refine the domain. A relational algebra formally defines the mapping metamodel, i.e. the abstract syntax of the mapping model. We also use the algebra to validate specific mapping models, i.e. to check whether they are valid instances of the mapping metamodel.

4.1.1 Mapping Patterns

The refinement of the domain extends the knowledge about the domain by mapping its representation from a higher layer to a more detailed representation at the subsequent lower layer. During the refinement, the representation of a higher-layer object is likely to change. We express the possible structural changes with a set of mapping patterns. These patterns define how the lower-layer representation of objects is derived from their higher-layer representation. The available mapping patterns are called *identity*, *replacement*, *merge*, *split*, *erasure*, and *appearance* and are illustrated in figure 4.3. We explain each pattern in the following.

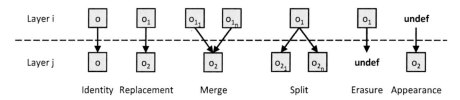

Figure 4.3: Mapping patterns

- The identity pattern leaves a higher-level object unchanged at the lower layer. The object o at the higher layer i has exactly the same representation at the lower layer j.

- The replacement pattern maps a higher-level object to a lower-level one. The object o_1 at layer i is represented by the object o_2 at layer j.

- The merge pattern maps multiple higher-level objects to one lower-level object. The objects o_{1_1} to o_{1_n} at layer i are represented by the object o_2 at layer j.

- The split pattern maps a higher-level object to one out of some lower-level objects. The object o_1 at layer i is represented by one out of the objects o_{2_1} to o_{2_n} at layer j. This pattern represents a choice between different mapping alternatives.

- It may be the case that a higher-level object is no more relevant at a lower level. For this purpose, the erasure pattern maps an object o_1 at layer i to no object at layer j, so that the object does not have any lower-level representation.

- The other way round, it may be the case that a new object o_2 is introduced at layer j, i.e. does not have any representation at layer i and thus cannot be derived from any other objects. This case is considered by the appearance pattern.

We establish a specific refinement of the domain from a higher layer to a lower layer by applying the refinement patterns to the objects of the domain model. An applied pattern is simply called a mapping. We use these mappings later for the automated refinement of policies that use the respective objects.

4.1.2 Mapping Metamodel

The mapping metamodel allows to specify how the domain is refined by mapping its representation from a higher to the subsequent lower layer, as shown in figure 4.4. The metamodel considers the presented mapping patterns and represents the abstract syntax of the mappings, i.e. it defines the structure of the mapping model.

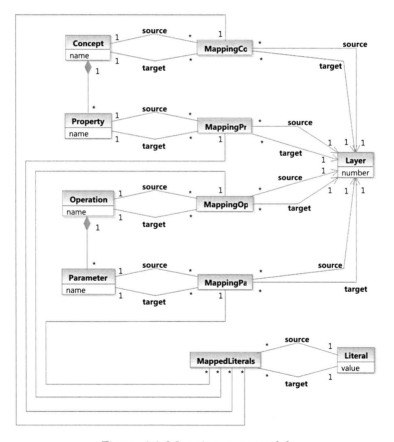

Figure 4.4: Mapping metamodel

The mapping metamodel performs the mapping as follows. Each mapping relates a source and a target layer with a source and a target object of the same entity. In the metamodel, only those objects appear that can be used by policies, i.e. mappings can be specified for concepts (*MappingCo*), properties (*MappingPr*), operations (*MappingOp*), and parameters (*MappingPa*). A concept mapping maps a source concept from a higher layer to a target concept at a lower layer, and the same applies for property, operation, and parameter mappings.

A mapping can optionally specify a literal that represents the source object and a literal that represents the target object. This will later be helpful in cases where objects are used in connection with literals. One example is the passing of a literal to a parameter of an invoked operation. Depending on the passed literal at the higher layer, one might pass a different literal when invoking the refined operation at the lower layer. Another example is the comparison of a property with a literal. Depending on the particular property and literal, a different comparison might be necessary at the lower layer.

In figure 4.4, it seems at first view that the mapping metamodel only considers the replacement pattern. In fact, the other mapping patterns do not have an explicit representation in the metamodel, but are considered special types or extensions of the replacement pattern, as described in the following.

4.1.3 Mapping Model

The mapping model specifies the mappings that refine the domain model from a higher layer to a subsequent lower layer. The developer can decide which objects to refine. It is not necessary to specify a complete refinement of the domain model that covers all available objects. However, the mapping model should specify the refinement of those objects used in the policies to be refined as these mappings represent the basis for the generation of refined policies. As the mapping model is an instance of the mapping metamodel, various restrictions must hold on its structure.

In the following, we formally define the abstract syntax of the mapping model with a relational algebra. Figure 4.5 shows an overview of the respective relations as an Entity Relationship (ER) diagram.

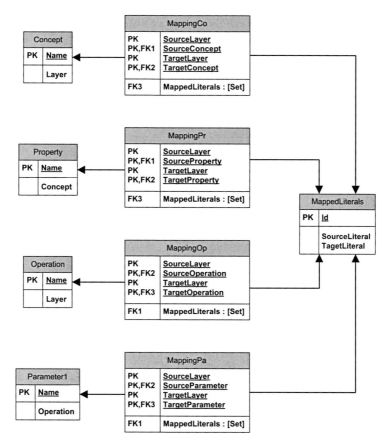

Figure 4.5: ER diagram of the mapping model

Definition 21 (*Mapping model*)

Let $MappingCo$ be a relation of concept mappings, $MappingPr$ a relation of property mappings, $MappingOp$ a relation of operation mappings, $MappingPa$ a relation of parameter mappings.

The mapping model

$$MM := (MappingCo, MappingPr, MappingOp, MappingPa) \qquad (4.1)$$

is a quadruple of relations that specifies the mappings to refine the concepts, properties, operations, and parameters of a domain model. A mapping model is valid if (4.2)–(4.6), (4.8)–(4.13), (4.15)–(4.19), and (4.21)–(4.26) hold for the enclosed relations.

Identity and Replacement

We now formally define identity and replacement mappings with a relational algebra. Identity and replacement mappings are defined in common as identity is regarded as a special form of replacement in which the source and target objects and the respective literals are the same. We perform the formalization with four relations, as illustrated in figure 4.5. Each relation considers a specific entity. $MappingCo$ defines identity and replacement mappings for concepts (cf. definition 22), $MappingPr$ for properties (cf. definition 23), $MappingOp$ for operations (cf. definition 24), and $MappingPa$ for parameters (cf. definition 25).

An identity or replacement mapping for concepts maps one source concept at a source layer to one target concept at a target layer (4.2). The respective concepts are indicated with their names. An undefined object $undef$ is used for different mapping patterns later. Optionally, literals that represent the source concept can be mapped to literals that represent the target concept (4.2,4.7). An alternative notation is provided for concept mappings (4.3,4.4). Source and target concepts can be the same (identity mapping) or different ones (replacement mapping). A concept mapping is only allowed from a higher layer to the subsequent lower layer (4.5). The source concept must be available at the source layer and the target concept at the target layer (4.6).

We define identity and replacement mappings for properties, operations, and parameters in the same way as for concepts. Additionally, a property mapping implies that a respective concept mapping is modeled, and a parameter mapping implies that a respective operation mapping is modeled. In the following, we provide the formal definitions of identity and replacement mappings.

Definition 22 (*Concept identity and replacement*)

Let Lit be the set of literals. Let Id_{La} be the set of layer identifiers, Id_{Co} the set of concept identifiers. Let $lit_1, lit_2 \in Lit$, $ml, ml_1, ml_2 \in \mathcal{P}(Lit \times Lit)$, $la_1, la_2 \in Id_{La}$, $co_1, co_2 \in Id_{Co}$.

The relation

$$MappingCo(SourceLayer, SourceConcept,$$
$$TargetLayer, TargetConcept, MappedLiterals) \qquad (4.2)$$
$$\subseteq Id_{La} \times Id_{Co} \cup \{undef\} \times Id_{La} \times Id_{Co} \cup \{undef\} \times \mathcal{P}(Lit \times Lit)$$

specifies which identity and replacement mappings for concepts are modeled.

The notation

$$co_1 \xrightarrow[la_1 \to la_2]{Co} co_2 :\Leftrightarrow \exists ml.(la_1, co_1, la_2, co_2, ml) \in MappingCo \qquad (4.3)$$

$$co_1[lit_1] \xrightarrow[la_1 \to la_2]{Co} co_2[lit_2]$$
$$\qquad\qquad (4.4)$$
$$:\Leftrightarrow \exists ml.(la_1, co_1, la_2, co_2, ml) \in MappingCo \wedge (lit_1, lit_2) \in ml$$

allows to specify identity and replacement mappings for concepts in an alternative way.

For concept mappings, the following must hold:

$$co_1 \xrightarrow[la_1 \to la_2]{Co} co_2 \in MappingCo \Rightarrow la_1 = la_2 - 1 \qquad (4.5)$$

$$co_1 \xrightarrow[la_1 \to la_2]{Co} co_2 \Rightarrow co_1 \in conceptsOfLayer(la_1) \cup \{undef\}$$
$$\qquad\qquad (4.6)$$
$$\wedge\, co_2 \in conceptsOfLayer(la_2) \cup \{undef\}$$

The function

$$mappedLiterals^{Co}_{la_1 \to la_2} : Id_{Co} \times Lit \times Id_{Co} \to \mathcal{P}(Lit)$$
$$(co_1, lit_1, co_2) \mapsto \{lit_2 | \exists ml.(la_1, co_1, la_2, co2, ml) \qquad (4.7)$$
$$\in MappingCo \wedge (lit_1, lit_2) \in ml\}$$

determines the literals of the target concept to which a source concept and literal are mapped.

Definition 23 (*Property identity and replacement*)

Let *Lit* be the set of literals. Let Id_{La} be the set of layer identifiers, Id_{Pr} the set of property identifiers. Let $lit_1, lit_2 \in Lit$, $ml, ml_1, ml_2 \in \mathcal{P}(Lit \times Lit)$, $la_1, la_2 \in Id_{La}$, $pr_1, pr_2 \in Id_{Pr}$.

The relation

$$MappingPr(SourceLayer, SourceProperty,$$
$$\quad TargetLayer, TargetProperty, MappedLiterals) \tag{4.8}$$
$$\quad \subseteq Id_{La} \times Id_{Pr} \cup \{undef\} \times Id_{La} \times Id_{Pr} \cup \{undef\} \times \mathcal{P}(Lit \times Lit)$$

specifies which identity and replacement mappings for properties are modeled.

The notation

$$pr_1 \xrightarrow[la_1 \to la_2]{Pr} pr_2 :\Leftrightarrow \exists ml.(la_1, pr_1, la_2, pr_2, ml) \in MappingPr \tag{4.9}$$

$$pr_1[lit_1] \xrightarrow[la_1 \to la_2]{Pr} pr_2[lit_2]$$
$$\quad :\Leftrightarrow \exists ml.(la_1, pr_1, la_2, pr_2, ml) \in MappingPr \wedge (lit_1, lit_2) \in ml \tag{4.10}$$

allows to specify identity and replacement mappings for properties in an alternative way.

For property mappings, the following must hold:

$$pr_1 \xrightarrow[la_1 \to la_2]{Pr} pr_2 \Rightarrow la_1 = la_2 - 1 \tag{4.11}$$

$$pr_1 \xrightarrow[la_1 \to la_2]{Pr} pr_2 \Rightarrow pr_1 \in propertiesOfLayer(la_1) \cup \{undef\}$$
$$\quad \wedge pr_2 \in propertiesOfLayer(la_2) \cup \{undef\} \tag{4.12}$$

$$pr_1 \xrightarrow[la_1 \to la_2]{Pr} pr_2 \Rightarrow co_1 \xrightarrow[la_1 \to la_2]{Co} co_2$$
$$\quad \text{where } co_1 = conceptOfProperty(pr_1) \tag{4.13}$$
$$\quad \wedge co_2 = conceptOfProperty(pr_2)$$

The function

$$mappedLiterals^{Pr}_{la_1 \to la_2} : Id_{Pr} \times Lit \times Id_{Pr} \to \mathcal{P}(Lit)$$
$$(pr_1, lit_1, pr_2) \mapsto \{lit_2 | \exists ml.(la_1, pr_1, la_2, pr_2, ml) \tag{4.14}$$
$$\quad \in MappingPr \wedge (lit_1, lit_2) \in ml\}$$

determines the literals of the target property to which a source property and literal are mapped.

Definition 24 (*Operation identity and replacement*)

Let Lit be the set of literals. Let Id_{La} be the set of layer identifiers, Id_{Op} the set of operation identifiers. Let $lit_1, lit_2 \in Lit$, $ml, ml_1, ml_2 \in \mathcal{P}(Lit \times Lit)$, $la_1, la_2 \in Id_{La}$, $op_1, op_2 \in Id_{Op}$.

The relation

$$MappingOp(SourceLayer, SourceOperation, \\ TargetLayer, TargetOperation, MappedLiterals) \qquad (4.15) \\ \subseteq Id_{La} \times Id_{Op} \cup \{undef\} \times Id_{La} \times Id_{Op} \cup \{undef\} \times \mathcal{P}(Lit \times Lit)$$

specifies which identity and replacement mappings for operations are modeled.

The notation

$$op_1 \xrightarrow[la_1 \to la_2]{Op} op_2 :\Leftrightarrow \exists ml.(la_1, op_1, la_2, op_2, ml) \in MappingOp \qquad (4.16)$$

$$op_1[lit_1] \xrightarrow[la_1 \to la_2]{Op} op_2[lit_2] \qquad (4.17) \\ :\Leftrightarrow \exists ml.(la_1, op_1, la_2, op_2, ml) \in MappingOp \wedge (lit_1, lit_2) \in ml$$

allows to specify identity and replacement mappings for operations in an alternative way.

For operation mappings, the following must hold:

$$op_1 \xrightarrow[la_1 \to la_2]{Op} op_2 \Rightarrow la_1 = la_2 - 1 \qquad (4.18)$$

$$op_1 \xrightarrow[la_1 \to la_2]{Op} op_2 \Rightarrow op_1 \in operationsOfLayer(la_1) \cup \{undef\} \qquad (4.19) \\ \wedge op_2 \in operationsOfLayer(la_2) \cup \{undef\}$$

The function

$$mappedLiterals_{la_1 \to la_2}^{Op} : Id_{Op} \times Lit \times Id_{Op} \to \mathcal{P}(Lit) \\ (op_1, lit_1, op_2) \mapsto \{lit_2 | \exists ml.(la_1, op_1, la_2, op_2, ml) \qquad (4.20) \\ \in MappingOp \wedge (lit_1, lit_2) \in ml\}$$

determines the literals of the target operation to which a source operation and literal are mapped.

Definition 25 (*Parameter identity and replacement*)

Let Lit be the set of literals. Let Id_{La} be the set of layer identifiers, Id_{Pa} the set of parameter identifiers. Let $lit_1, lit_2 \in Lit$, $ml, ml_1, ml_2 \in \mathcal{P}(Lit \times Lit)$, $la_1, la_2 \in Id_{La}$, $pa_1, pa_2 \in Id_{Pa}$.

The relation

$$MappingPa(SourceLayer, SourceParameter,$$
$$TargetLayer, TargetParameter, MappedLiterals) \tag{4.21}$$
$$\subseteq Id_{La} \times Id_{Pa} \cup \{undef\} \times Id_{La} \times Id_{Pa} \cup \{undef\} \times \mathcal{P}(Lit \times Lit)$$

specifies which identity and replacement mappings for operations are modeled.

The notation

$$pa_1 \xrightarrow[la_1 \to la_2]{Pa} pa_2 :\Leftrightarrow \exists ml.(la_1, pa_1, la_2, pa_2, ml) \in MappingPa \tag{4.22}$$

$$pa_1[lit_1] \xrightarrow[la_1 \to la_2]{Pa} pa_2[lit_2] \tag{4.23}$$
$$:\Leftrightarrow \exists ml.(la_1, pa_1, la_2, pa_2, ml) \in MappingPa \wedge (lit_1, lit_2) \in ml$$

allows to specify the identity and replacement mappings for parameters in an alternative way.

For parameter mappings, the following must hold:

$$pa_1 \xrightarrow[la_1 \to la_2]{Pa} pa_2 \Rightarrow la_1 = la_2 - 1 \tag{4.24}$$

$$pa_1 \xrightarrow[la_1 \to la_2]{Pa} pa_2 \Rightarrow pa_1 \in parametersOfLayer(la_1) \cup \{undef\}$$
$$\wedge\, pa_2 \in parametersOfLayer(la_2) \cup \{undef\} \tag{4.25}$$

$$pa_1 \xrightarrow[la_1 \to la_2]{Pa} pa_2 \Rightarrow op_1 \xrightarrow[la_1 \to la_2]{Op} op_2$$
$$\text{where } op_1 = operationOfParameter(pa_1)$$
$$\wedge\, op_2 = operationOfParameter(pa_2) \tag{4.26}$$

The function

$$mappedLiterals^{Pa}_{la_1 \to la_2} : Id_{Pa} \times Lit \times Id_{Pa} \to \mathcal{P}(Lit)$$
$$(pa_1, lit_1, pa_2) \mapsto \{lit_2 | \exists ml.(la_1, pa_1, la_2, pa_2, ml) \tag{4.27}$$
$$\in MappingPa \wedge (lit_1, lit_2) \in ml\}$$

determines the literals of the target parameter to which a source parameter and literal are mapped.

Merge

Now, we formally define merge mappings with a relational algebra. We regard a merge mapping as a combination of replacement mappings that have the same target object, i.e. different higher-layer objects are replaced with the same lower-layer object. Thus, no distinct relation is necessary for the merge pattern, and we perform the formalization with the four relations, as illustrated in figure 4.5. This is a convenient way to keep the formalization simple.

Definition 26 (*Concept merge*)

Let Lit be the set of literals. Let Id_{La} be the set of layer identifiers, Id_{Co} the set of concept identifiers. Let $lit_{1_1}, ..., lit_{1_n}, lit_2 \in Lit$, $la_1, la_2 \in Id_{La}$, $co_{1_1}, ..., co_{1_n}, co_2 \in Id_{Co}$.

The notation

$$(co_{1_1}, ..., co_{1_n}) \xrightarrow[la_1 \to la_2]{Co} co_2 :\Leftrightarrow \forall i \in \{1, ..., n\}.co_{1_i} \xrightarrow[la_1 \to la_2]{Co} co_2 \tag{4.28}$$

$$(co_{1_1}[lit_{1_1}], ..., co_{1_n}[lit_{1_n}]) \xrightarrow[la_1 \to la_2]{Co} co_2[lit_2]$$

$$:\Leftrightarrow \forall i \in \{1, ..., n\}.co_{1_i}[lit_{1_i}] \xrightarrow[la_1 \to la_2]{Co} co_2[lit_2] \tag{4.29}$$

allows to specify merge mappings for concepts.

Definition 27 (*Property merge*)

Let Lit be the set of literals. Let Id_{La} be the set of layer identifiers, Id_{Pr} the set of property identifiers. Let $lit_{1_1}, ..., lit_{1_n}, lit_2 \in Lit$, $la_1, la_2 \in Id_{La}$, $pr_{1_1}, ..., pr_{1_n}, pr_2 \in Id_{Pr}$.

The notation

$$(pr_{1_1}, ..., pr_{1_n}) \xrightarrow[la_1 \to la_2]{Pr} pr_2 :\Leftrightarrow \forall i \in \{1, ..., n\}.pr_{1_i} \xrightarrow[la_1 \to la_2]{Pr} pr_2 \tag{4.30}$$

$$(pr_{1_1}[lit_{1_1}], ..., pr_{1_n}[lit_{1_n}]) \xrightarrow[la_1 \to la_2]{Pr} pr_2[lit_2]$$

$$:\Leftrightarrow \forall i \in \{1, ..., n\}.pr_{1_i}[lit_{1_i}] \xrightarrow[la_1 \to la_2]{Pr} pr_2[lit_2] \tag{4.31}$$

allows to specify merge mappings for properties.

Definition 28 (*Operation merge*)

Let *Lit* be the set of literals. Let Id_{La} be the set of layer identifiers, Id_{Op} the set of operation identifiers. Let $lit_{1_1}, ..., lit_{1_n}, lit_2 \in Lit$, $la_1, la_2 \in Id_{La}$, $op_{1_1}, ..., op_{1_n}, op_2 \in Id_{Op}$.

The notation

$$(op_{1_1}, ..., op_{1_n}) \xrightarrow[la_1 \to la_2]{Op} op_2 :\Leftrightarrow \forall i \in \{1, ..., n\}.op_{1_i} \xrightarrow[la_1 \to la_2]{Op} op_2 \qquad (4.32)$$

$$(op_{1_1}[lit_{1_1}], ..., op_{1_n}[lit_{1_n}]) \xrightarrow[la_1 \to la_2]{Op} op_2[lit_2]$$

$$:\Leftrightarrow \forall i \in \{1, ..., n\}.op_{1_i}[lit_{1_i}] \xrightarrow[la_1 \to la_2]{Op} op_2[lit_2] \qquad (4.33)$$

allows to specify merge mappings for operations.

Definition 29 (*Parameter merge*)

Let *Lit* be the set of literals. Let Id_{La} be the set of layer identifiers, Id_{Pa} the set of parameter identifiers. Let $lit_{1_1}, ..., lit_{1_n}, lit_2 \in Lit$, $la_1, la_2 \in Id_{La}$, $pa_{1_1}, ..., pa_{1_n}, pa_2 \in Id_{Pa}$.

The notation

$$(pa_{1_1}, ..., pa_{1_n}) \xrightarrow[la_1 \to la_2]{Pa} pa_2 :\Leftrightarrow \forall i \in \{1, ..., n\}.pa_{1_i} \xrightarrow[la_1 \to la_2]{Pa} pa_2 \qquad (4.34)$$

$$(pa_{1_1}[lit_{1_1}], ..., pa_{1_n}[lit_{1_n}]) \xrightarrow[la_1 \to la_2]{Pa} pa_2[lit_2]$$

$$:\Leftrightarrow \forall i \in \{1, ..., n\}.pa_{1_i}[lit_{1_i}] \xrightarrow[la_1 \to la_2]{Pa} pa_2[lit_2] \qquad (4.35)$$

allows to specify merge mappings for parameters.

Split

We now formally define split mappings with a relational algebra. Similarly to merge patterns, we regard a split mapping as a combination of replacement mappings that have the same source object, i.e. the same higher-layer object is replaced with different lower-layer objects. Thus, no distinct relation is necessary for the split pattern, and we perform the formalization with the four relations, as illustrated in figure 4.5. This further helps to keep the formalization simple.

Definition 30 (*Concept split*)

Let *Lit* be the set of literals. Let Id_{La} be the set of layer identifiers, Id_{Co} the set of concept identifiers. Let $lit_{1_1}, ..., lit_{1_n}, lit_2 \in Lit$, $la_1, la_2 \in Id_{La}$, $co_1, co_{2_1}, ..., co_{2_n} \in Id_{Co}$.

The notation

$$co_1 \xrightarrow[la_1 \to la_2]{Co} (co_{2_1}, ..., co_{2_n}) :\Leftrightarrow \forall i \in \{1, ..., n\}.co_1 \xrightarrow[la_1 \to la_2]{Co} co_{2_i} \qquad (4.36)$$

$$co_1[lit_1] \xrightarrow[la_1 \to la_2]{Co} (co_{2_1}[lit_{2_1}], ..., co_{2_n}[lit_{2_n}])$$
$$:\Leftrightarrow \forall i \in \{1, ..., n\}.co_1[lit_1] \xrightarrow[la_1 \to la_2]{Co} co_{2_i}[lit_{2_i}] \qquad (4.37)$$

allows to specify split mappings for concepts.

Definition 31 (*Property split*)

Let *Lit* be the set of literals. Let Id_{La} be the set of layer identifiers, Id_{Pr} the set of property identifiers. Let $lit_{1_1}, ..., lit_{1_n}, lit_2 \in Lit$, $la_1, la_2 \in Id_{La}$, $pr_1, ..., pr_{2_1}, pr_{2_n} \in Id_{Pr}$.

The notation

$$pr_1 \xrightarrow[la_1 \to la_2]{Pr} (pr_{2_1}, ..., pr_{2_n}) :\Leftrightarrow \forall i \in \{1, ..., n\}.pr_1 \xrightarrow[la_1 \to la_2]{Pr} pr_{2_i} \qquad (4.38)$$

$$pr_1[lit_1] \xrightarrow[la_1 \to la_2]{Pr} (pr_{2_1}[lit_{2_1}], ..., pr_{2_n}[lit_{2_n}])$$
$$:\Leftrightarrow \forall i \in \{1, ..., n\}.pr_1[lit_1] \xrightarrow[la_1 \to la_2]{Pr} pr_{2_i}[lit_{2_i}] \qquad (4.39)$$

allows to specify split mappings for properties.

Definition 32 (*Operation split*)

Let *Lit* be the set of literals. Let Id_{La} be the set of layer identifiers, Id_{Op} the set of operation identifiers. Let $lit_{1_1}, ..., lit_{1_n}, lit_2 \in Lit$, $la_1, la_2 \in Id_{La}$, $op_1, ..., op_{2_1}, op_{2_n} \in Id_{Op}$.

The notation

$$op_1 \xrightarrow[la_1 \to la_2]{Op} (op_{2_1}, ..., op_{2_n}) :\Leftrightarrow \forall i \in \{1, ..., n\}.op_1 \xrightarrow[la_1 \to la_2]{Op} op_{2_i} \qquad (4.40)$$

$$op_1[lit_1] \xrightarrow[la_1 \to la_2]{Op} (op_{2_1}[lit_{2_1}], ..., op_{2_n}[lit_{2_n}])$$
$$:\Leftrightarrow \forall i \in \{1, ..., n\}.op_1[lit_1] \xrightarrow[la_1 \to la_2]{Op} op_{2_i}[lit_{2_i}] \qquad (4.41)$$

allows to specify split mappings for operations.

Definition 33 (*Parameter split*)

Let *Lit* be the set of literals. Let Id_{La} be the set of layer identifiers, Id_{Pa} the set of parameter identifiers. Let $lit_1, ..., lit_{1_n}, lit_2 \in Lit, la_1, la_2 \in Id_{La}$, $pa_1, ..., pa_{2_1}, pa_{2_n} \in Id_{Pa}$.

The notation

$$pa_1 \xrightarrow[la_1 \to la_2]{Pa} (pa_{2_1}, ..., pa_{2_n}) :\Leftrightarrow \forall i \in \{1, ..., n\}.pa_1 \xrightarrow[la_1 \to la_2]{Pa} pa_{2_i} \qquad (4.42)$$

$$pa_1[lit_1] \xrightarrow[la_1 \to la_2]{Pa} (pa_{2_1}[lit_{2_1}], ..., pa_{2_n}[lit_{2_n}])$$
$$:\Leftrightarrow \forall i \in \{1, ..., n\}.pa_1[lit_1] \xrightarrow[la_1 \to la_2]{Pa} pa_{2_i}[lit_{2_i}] \qquad (4.43)$$

allows to specify split mappings for parameters.

Erasure

Now, we formally define erasure mappings with a relational algebra. We regard an erasure mapping as a special form of a replacement mapping. If an object does not have a lower-layer representation, it is replaced with the special element *undef*. Thus, no distinct relation is necessary for the erasure pattern, and we perform the formalization with the four relations, as illustrated in figure 4.5. This further helps to keep the formalization simple.

Definition 34 (*Concept erasure*)

Let *Lit* be the set of literals. Let Id_{La} be the set of layer identifiers, Id_{Co} the set of concept identifiers. Let $lit_1 \in Lit, ml \in \mathcal{P}(Lit \times Lit)$, $la_1, la_2 \in Id_{La}, co_1 \in Id_{Co}$.

The notation

$$co_1 \xrightarrow[la_1 \to la_2]{Co} :\Leftrightarrow \exists ml.(la_1, co_1, la_2, undef, ml) \in MappingCo \qquad (4.44)$$

$$co_1[lit_1] \xrightarrow[la_1 \to la_2]{Co} :\Leftrightarrow \exists ml.$$
$$(la_1, co_1, la_2, undef, ml) \in MappingCo \wedge (lit_1, undef) \in ml \qquad (4.45)$$

allows to specify erasure mappings for concepts in an alternative way.

The function

$$erasedLiterals^{Co}_{la_1 \to la_2} : Id_{Co} \to \mathcal{P}(Lit)$$
$$co_1 \mapsto \{lit_1 | \exists ml.(la_1, co_1, la_2, undef, ml) \in MappingCo \qquad (4.46)$$
$$\wedge (lit_1, undef) \in ml\}$$

determines the literals of the source concept that are erased with it.

Definition 35 (*Property erasure*)

Let *Lit* be the set of literals. Let Id_{La} be the set of layer identifiers, Id_{Pr} the set of property identifiers. Let $lit_1 \in Lit$, $ml \in \mathcal{P}(Lit \times Lit)$, $la_1, la_2 \in Id_{La}$, $pr_1 \in Id_{Pr}$.

The notation

$$pr_1 \xrightarrow[la_1 \to la_2]{Pr} :\Leftrightarrow \exists ml.(la_1, pr_1, la_2, undef, ml) \in MappingPr \qquad (4.47)$$

$$pr_1[lit_1] \xrightarrow[la_1 \to la_2]{Pr} :\Leftrightarrow \exists ml.$$
$$(la_1, pr_1, la_2, undef, ml) \in MappingPr \wedge (lit_1, undef) \in ml \qquad (4.48)$$

allows to specify erasure mappings for properties in an alternative way.

The function

$$erasedLiterals^{Pr}_{la_1 \to la_2} : Id_{Pr} \to \mathcal{P}(Lit)$$
$$pr_1 \mapsto \{lit_1 | \exists ml.(la_1, pr_1, la_2, undef, ml) \in MappingPr \qquad (4.49)$$
$$\wedge (lit_1, undef) \in ml\}$$

determines the literals of the source property that are erased with it.

Definition 36 (*Operation erasure*)

Let Lit be the set of literals. Let Id_{La} be the set of layer identifiers, Id_{Op} the set of operation identifiers. Let $lit_1 \in Lit$, $ml \in \mathcal{P}(Lit \times Lit)$, $la_1, la_2 \in Id_{La}$, $op_1 \in Id_{Op}$.

The notation

$$op_1 \xrightarrow[la_1 \rightarrow la_2]{Op} :\Leftrightarrow \exists ml.(la_1, op_1, la_2, undef, ml) \in MappingOp \tag{4.50}$$

$$op_1[lit_1] \xrightarrow[la_1 \rightarrow la_2]{Op} :\Leftrightarrow \exists ml.$$
$$(la_1, op_1, la_2, undef, ml) \in MappingOp \wedge (lit_1, undef) \in ml \tag{4.51}$$

allows to specify erasure mappings for operations in an alternative way.

The function

$$erasedLiterals_{la_1 \rightarrow la_2}^{Op} : Id_{Op} \rightarrow \mathcal{P}(Lit)$$
$$op_1 \mapsto \{lit_1 | \exists ml.(la_1, op_1, la_2, undef, ml) \in MappingOp \tag{4.52}$$
$$\wedge (lit_1, undef) \in ml\}$$

determines the literals of the source operation that are erased with it.

Definition 37 (*Parameter erasure*)

Let Lit be the set of literals. Let Id_{La} be the set of layer identifiers, Id_{Pa} the set of parameter identifiers. Let $lit_1 \in Lit$, $ml \in \mathcal{P}(Lit \times Lit)$, $la_1, la_2 \in Id_{La}$, $pa_1 \in Id_{Pa}$.

The notation

$$pa_1 \xrightarrow[la_1 \rightarrow la_2]{Pa} :\Leftrightarrow \exists ml.(la_1, pa_1, la_2, undef, ml) \in MappingPa \tag{4.53}$$

$$pa_1[lit_1] \xrightarrow[la_1 \rightarrow la_2]{Pa} :\Leftrightarrow \exists ml.$$
$$(la_1, pa_1, la_2, undef, ml) \in MappingPa \wedge (lit_1, undef) \in ml \tag{4.54}$$

allows to specify erasure mappings for parameters in an alternative way.

The function

$$erasedLiterals^{Pa}_{la_1 \to la_2} : Id_{Pa} \to \mathcal{P}(Lit)$$
$$pa_1 \mapsto \{lit_1 | \exists ml.(la_1, pa_1, la_2, undef, ml) \in MappingPa \qquad (4.55)$$
$$\wedge (lit_1, undef) \in ml\}$$

determines the literals of the source parameter that are erased with it.

Appearance

We now formally define appearance mappings with a relational algebra. Similarly to erasure mappings, we regard an appearance mapping as a special form of a replacement mapping. If an object does not have a higher-layer representation, it is replaced from the special element *undef*. Thus, no distinct relation is necessary for the appearance pattern, and we perform the formalization with the four relations, as illustrated in figure 4.5. This further helps to keep the formalization simple.

Definition 38 (*Concept appearance*)

Let *Lit* be the set of literals. Let Id_{La} be the set of layer identifiers, Id_{Co} the set of concept identifiers. Let $lit_2 \in Lit$, $ml \in \mathcal{P}(Lit \times Lit)$, $la_1, la_2 \in Id_{La}$, $co_2 \in Id_{Co}$.

The notation

$$\xrightarrow[la_1 \to la_2]{Co} co_2 :\Leftrightarrow \exists ml.(la_1, undef, la_2, co_2, ml) \in MappingCo \qquad (4.56)$$

$$\xrightarrow[la_1 \to la_2]{Co} co_2[lit_2] :\Leftrightarrow \exists ml. \qquad (4.57)$$
$$(la_1, undef, la_2, co_2, ml) \in MappingCo \wedge (undef, lit_2) \in ml$$

specifies which appearance mappings for concepts are modeled.

The function

$$appearedLiterals^{Co}_{la_1 \to la_2} : Id_{Co} \to \mathcal{P}(Lit)$$
$$co_2 \mapsto \{lit_2 | \exists ml.(la_1, undef, la_2, co_2, ml) \in MappingCo \qquad (4.58)$$
$$\wedge (undef, lit_2) \in ml\}$$

determines the literals of the target concept that appear with it.

Definition 39 (*Property appearance*)

Let *Lit* be the set of literals. Let Id_{La} be the set of layer identifiers, Id_{Pr} the set of property identifiers. Let $lit_2 \in Lit$, $ml \in \mathcal{P}(Lit \times Lit)$, $la_1, la_2 \in Id_{La}$, $pr_2 \in Id_{Pr}$.

The notation

$$\xrightarrow[la_1 \to la_2]{Pr} pr_2 :\Leftrightarrow \exists ml.(la_1, undef, la_2, pr_2, ml) \in MappingPr \qquad (4.59)$$

$$\xrightarrow[la_1 \to la_2]{Pr} pr_2[lit_2] :\Leftrightarrow \exists ml. \qquad (4.60)$$
$$(la_1, undef, la_2, pr_2, ml) \in MappingPr \wedge (undef, lit_2) \in ml$$

specifies which appearance mappings for properties are modeled.

The function

$$appearedLiterals^{Pr}_{la_1 \to la_2} : Id_{Pr} \to \mathcal{P}(Lit)$$
$$pr_2 \mapsto \{lit_2 | \exists ml.(la_1, undef, la_2, pr_2, ml) \in MappingPr \qquad (4.61)$$
$$\wedge (undef, lit_2) \in ml\}$$

determines the literals of the target property that appear with it.

Definition 40 (*Operation appearance*)

Let *Lit* be the set of literals. Let Id_{La} be the set of layer identifiers, Id_{Op} the set of operation identifiers. Let $lit_2 \in Lit$, $ml \in \mathcal{P}(Lit \times Lit)$, $la_1, la_2 \in Id_{La}$, $op_2 \in Id_{Op}$.

The notation

$$\xrightarrow[la_1 \to la_2]{Op} op_2 :\Leftrightarrow \exists ml.(la_1, undef, la_2, op_2, ml) \in MappingOp \qquad (4.62)$$

$$\xrightarrow[la_1 \to la_2]{Op} op_2[lit_2] :\Leftrightarrow \exists ml. \qquad (4.63)$$
$$(la_1, undef, la_2, op_2, ml) \in MappingOp \wedge (undef, lit_2) \in ml$$

specifies which appearance mappings for operations are modeled.

The function

$$appearedLiterals^{Op}_{la_1 \to la_2} : Id_{Op} \to \mathcal{P}(Lit)$$
$$op_2 \mapsto \{lit_2 | \exists ml.(la_1, undef, la_2, op_2, ml) \in MappingOp \qquad (4.64)$$
$$\wedge (undef, lit_2) \in ml\}$$

determines the literals of the target operation that appear with it.

Definition 41 (*Parameter appearance*)

Let *Lit* be the set of literals. Let Id_{La} be the set of layer identifiers, Id_{Pa} the set of parameter identifiers. Let $lit_2 \in Lit$, $ml \in \mathcal{P}(Lit \times Lit)$, $la_1, la_2 \in Id_{La}$, $pa_2 \in Id_{Pa}$.

The notation

$$\xrightarrow[la_1 \to la_2]{Pa} pa_2 :\Leftrightarrow \exists ml.(la_1, undef, la_2, pa_2, ml) \in MappingPa \qquad (4.65)$$

$$\xrightarrow[la_1 \to la_2]{Pa} pa_2[lit_2] :\Leftrightarrow \exists ml. \qquad (4.66)$$
$$(la_1, undef, la_2, pa_2, ml) \in MappingPa \wedge (undef, lit_2) \in ml$$

specifies which appearance mappings for parameters are modeled.

The function

$$appearedLiterals^{Pa}_{la_1 \to la_2} : Id_{Pa} \to \mathcal{P}(Lit)$$
$$pa_2 \mapsto \{lit_2 | \exists ml.(la_1, undef, la_2, pa_2, ml) \in MappingPa \qquad (4.67)$$
$$\wedge (undef, lit_2) \in ml\}$$

determines the literals of the target parameter that appear with it.

4.1.4 Example

In the following, we present a mapping model that refines the domain model provided in section 3.3.3. This mapping model provides mappings from the first to the second abstraction layer. Figure 4.6 shows an overview of the mapping model with a concrete syntax based on Unified Modeling Language (UML). Tables 4.1 to 4.4 provide the formal representation of the mapping model.

The mappings are rather straightforward. We replace the concept *signalQuality* and its properties with the concept *intensityChange* and its properties. In the same way, we replace the concept *cellPhone* and its property with the concept *device* and its property. Finally, we merge the two operations *increasePower* and *decreasePower* into one operation *changeTXP*. The parameter *ctChangeValue* appears newly at the second abstraction layer.

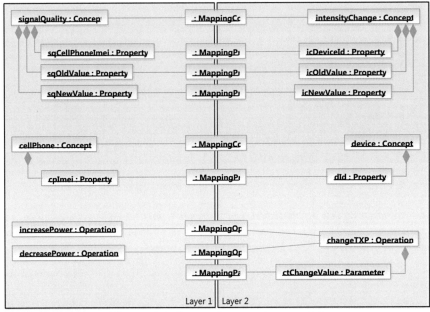

Figure 4.6: Example mapping model

Table 4.1: Relation *MappingCo*

SourceLayer	SourceConcept	TargetLayer	TargetConcept	MappedLiterals
1	signalQuality	2	intensityChange	∅
1	cellPhone	2	device	∅

Table 4.2: Relation *MappingPr*

SourceLayer	SourceProperty	TargetLayer	TargetProperty	MappedLiterals
1	sqCellPhoneImei	2	icDeviceId	∅
1	sqOldValue	2	icOldValue	∅
1	sqNewValue	2	icNewValue	∅
1	cpImei	2	dId	∅

Table 4.3: Relation *MappingOp*

SourceLayer	SourceOperation	TargetLayer	TargetOperation	MappedLiterals
1	increasePower	2	changeTXP	∅
1	decreasePower	2	changeTXP	∅

Table 4.4: Relation *MappingPa*

SourceLayer	SourceParameter	TargetLayer	TargetParameter	MappedLiterals
1	undef	2	ctChangeValue	{(undef,undef)}

4.2 Policy Refinement

The refinement of the domain from a higher to a lower layer described in the previous section represents the basis for the refinement of policies within this domain. For this purpose, knowledge about the refinement of the domain was formalized in the mapping model. Now, refinement rules apply this knowledge to the high-level policy and linking models and generate the refined low-level policy and linking models in an automated way.

The refinement process is illustrated in figure 4.7. This figure exemplarily shows two layers i and j, but all layers are respected for the generation of refined policies, of course. If more than two layers are available, refinement is an iterative process from the highest to the lowest layer. In each iteration, the policies are refined with respect to the specified mappings. At the lowest layer, the policies are finally represented in a machine-executable way and ready for deployment. From this technical representation of the models, executable code in a policy language can be generated, so that the policies can be executed by a respective policy engine.

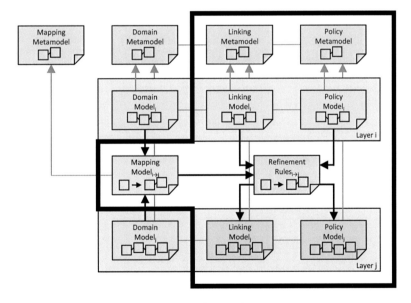

Figure 4.7: Policy refinement

In the following sections, we describe how we use refinement rules to generate refined policies. We formally define the refinement rules with a relational algebra and base this definition on the formal definition of the models. The refinement rules represent the formal semantics of the refinement process. We also prove that they preserve the static semantics of the models, i.e. that they refine valid models into valid models again.

4.2.1 Refinement Rules

We use refinement rules to refine the policies in the policy model and the links in the linking model from a higher layer into a lower layer. Refinement rules describe a Model-to-Model (M2M) transformation in a formal and language-independent way, as shown in figure 4.8. They are defined between the policy and linking metamodels as both source and target, and they transform the view of the policy and linking models at a higher layer into the view of the policy and linking models at a lower layer. The transformation extends the policy model with the refined policies and the linking model with the refined links.

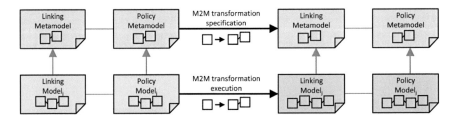

Figure 4.8: Refinement rule

A refinement rule is the smallest unit of transformation. It represents a declarative rewrite rule in a relational algebra that has a Left Hand Side (LHS) referring to some views as source, and that has a Right Hand Side (RHS) referring to some views as target. The LHS represents the input of the transformation, and the RHS represents the output of the transformation. We explain the structure of a refinement rule by means of (4.68), which is an example rule for the refinement of events.

$$
\left.
\begin{array}{l}
\forall la_1, la_2, po_1, po_2, ev_1. \\[2mm]
\left.
\begin{array}{l}
po_1 \xrightarrow[la_1 \to la_2]{Po} po_2 \quad \text{\Big\} Precondition} \\[3mm]
\wedge (po_1, ev_1) \in Event_{la_1} \quad \text{\} Filter}
\end{array}
\right\} \text{LHS} \\[6mm]
\Rightarrow \exists ev_2. \\[2mm]
\left.
\begin{array}{l}
(po_2, ev_2) \in Event_{la_2} \quad \text{\} Generator} \\[3mm]
\wedge ev_1 \xrightarrow[la_1 \to la_2]{Ev} ev_2 \quad \text{\} Postcondition}
\end{array}
\right\} \text{RHS}
\end{array}
\right. \tag{4.68}
$$

The LHS of a refinement rule consists of an optional precondition and a mandatory filter. The precondition can require that some other refinement was performed previously or some mapping was specified in the mapping model. Refinement rule (4.68) requires as a precondition that a policy po_1 at layer la_1 was refined into a policy po_2 at layer la_2. Besides that, the filter represents a pattern that describes a fragment of the policy or linking model. This pattern is matched with the respective relation to filter the desired tuples from it. Refinement rule (4.68) filters all tuples (po_1, ev_1) from the view $Event_{la_1}$.

The RHS of a refinement rule consists of a mandatory generator and an optional postcondition. The generator describes the refined fragment to be added to the policy or linking model. For all tuples that were matched by the filter of the LHS, the generator inserts refined tuples into the respective relations. In refinement rule (4.68), the refined tuple (po_2, ev_2) is inserted into the view $Event_{la_2}$, and this tuple associates a refined event ev_2 with the refined policy po_2. The name of the refined event ev_2 is not specified as the name is irrelevant for the formal semantics of the refinement process. It is only relevant that a refined event exists, but not how this event is named. An implementation would have to decide which name to choose. The postcondition is used to record tracing information. Refinement rule (4.68) records that ev_1 was refined into ev_2. This trace record might then trigger other refinement rules that have a respective precondition.

According to [169,170], the model transformation has the following features. It is unidirectional and executed in top-down direction with respect to the abstraction layers. The view of the policy and linking models at a higher layer serves as source, and the view at a lower layer serves as target. Refinement rules are applied as an endogenous transformation, i.e. the associated source and target metamodels are the same. Furthermore, refinement rules represent an in-place-transformation, i.e. the source and target models of the transformation are the same. The policy and linking models serve as input and are updated by the transformation. As they are both read and written, their static mode in the transformation is in/out. Although the transformation is an in-place one, it operates in extension-only mode, i.e. it does not remove any matched objects by the LHS from the linking model, but only adds the transformed objects to the model according to the RHS.

Tracing

As mentioned before, the refinement rules do not only filter and generate tuples, but they also generate trace records in their postcondition and refer to trace records in their precondition. We present the purposes and formal definition of the trace with a relational algebra in the following.

- The first purpose of the trace is to record which objects were refined into which other objects. The postcondition of refinement rule (4.68) e.g. records that the event ev_1 was refined into the event ev_2.

- The second purpose of the trace is to manage the control flow of the refinement process. Refinement rules with a precondition are triggered after an appropriate trace record was generated by another refinement rule. Refinement rule (4.68) e.g. is triggered after a policy po_1 was refined into another policy po_2.

- The third purpose of the trace is to ensure a unique refinement. The event ev_1 in refinement rule (4.68) e.g. might be assigned to multiple policies, and for each of them the rule will trigger the refinement into an event ev_2. Despite multiple occurrences of ev_1, the respective trace record ensures that only one event ev_2 is generated.

Definition 42 (*Trace*)

Let *TracePo* be a relation of trace records for policies, *TraceEv* a relation of trace records for events, *TraceCd* a relation of trace records for conditions, *TraceAc* a relation of trace records for actions, *TraceOI* a relation of trace records for operation invokings.

The trace

$$T := (TracePo, TraceEv, TraceCd, TraceAc, TraceOI) \tag{4.69}$$

is a quintuple that specifies which trace records were generated for policies, events, conditions, actions, and operation invokings.

Definition 42 refers to some relations that have not yet been defined, but are defined in the following. We perform the formalization with five relations in order to cover all relevant entities. *TracePo* defines the trace for policies (cf. definition 43), *TraceEv* for events (cf. definition 44), *TraceCd* for conditions (cf. definition 45), *TraceAc* for actions (cf. definition 46), and *TraceOI* for operation invokings (cf. definition 47).

We define trace records similarly to the mappings of the mapping model. A policy trace record records the refinement of a source policy at a source layer into a target policy at a target layer (4.70). The respective policies are indicated with their names. An alternative notation is provided for policy trace records (4.71). Any recorded refinement must be unambiguous (4.72). Also, two different policies must not be refined into the same policy (4.73). A policy trace record is only allowed from a higher layer to the subsequent lower layer (4.74). The source policy must be available at the source layer and the target policy at the target layer (4.75).

We define trace records for events, conditions, actions, and operation invokings in the same way as for policies. In the following, we provide their formal definitions.

Definition 43 (*Policy trace*)

Let Id_{La} be the set of layer identifiers, Id_{Po} the set of policy identifiers. Let $la_1, la_2 \in Id_{La}, po_1, po_2, po_3 \in Id_{Po}$.

The relation

$$TracePo(SourceLayer, SourcePolicy, TargetLayer, \\ TargetPolicy) \subseteq Id_{La} \times Id_{Po} \times Id_{La} \times Id_{Po} \tag{4.70}$$

specifies which trace records for policies were generated.

The notation

$$po_1 \xrightarrow[la_1 \to la_2]{Po} po_2 :\Leftrightarrow (la_1, po_1, la_2, po_2) \in TracePo \tag{4.71}$$

allows to specify trace records for policies in an alternative way.

For the policy trace, the following must hold:

$$po_1 \xrightarrow[la_1 \to la_2]{Po} po_2 \wedge po_1 \xrightarrow[la_1 \to la_2]{Po} po_3 \Rightarrow po_2 = po_3 \tag{4.72}$$

$$po_1 \xrightarrow[la_1 \to la_2]{Po} po_3 \wedge po_2 \xrightarrow[la_1 \to la_2]{Po} po_3 \Rightarrow po_1 = po_2 \tag{4.73}$$

$$po_1 \xrightarrow[la_1 \to la_2]{Po} po_2 \Rightarrow la_1 = la_2 - 1 \tag{4.74}$$

$$po_1 \xrightarrow[la_1 \to la_2]{Po} po_2 \Rightarrow po_1 \in policiesOfLayer(la_1) \\ \wedge po_2 \in policiesOfLayer(la_2) \tag{4.75}$$

Definition 44 (*Event trace*)

Let Id_{La} be the set of layer identifiers, Id_{Ev} the set of event identifiers. Let $la_1, la_2 \in Id_{La}$, $ev_1, ev_2, ev_3 \in Id_{Ev}$.

The relation

$$TraceEv(SourceLayer, SourceEvent, TargetLayer, \\ TargetEvent) \subseteq Id_{La} \times Id_{Ev} \times Id_{La} \times Id_{Ev} \tag{4.76}$$

specifies which trace records for events were generated.

The notation

$$ev_1 \xrightarrow[la_1 \to la_2]{Ev} ev_2 :\Leftrightarrow (la_1, ev_1, la_2, ev_2) \in TraceEv \tag{4.77}$$

allows to specify trace records for events in an alternative way.

For the event trace, the following must hold:

$$ev_1 \xrightarrow[la_1 \to la_2]{Ev} ev_2 \wedge ev_1 \xrightarrow[la_1 \to la_2]{Ev} ev_3 \Rightarrow ev_2 = ev_3 \tag{4.78}$$

$$ev_1 \xrightarrow[la_1 \to la_2]{Ev} ev_3 \in TraceEv \wedge ev_2 \xrightarrow[la_1 \to la_2]{Ev} ev_3 \Rightarrow ev_1 = ev_2 \tag{4.79}$$

$$ev_1 \xrightarrow[la_1 \to la_2]{Ev} ev_2 \in TraceEv \Rightarrow la_1 = la_2 - 1 \tag{4.80}$$

$$ev_1 \xrightarrow[la_1 \to la_2]{Ev} ev_2 \in TraceEv \Rightarrow ev_1 \in eventsOfLayer(la_1) \\ \wedge ev_2 \in eventsOfLayer(la_2) \tag{4.81}$$

Definition 45 (*Condition trace*)

Let Id_{La} be the set of layer identifiers, Id_{Co} the set of condition identifiers. Let $la_1, la_2 \in Id_{La}$, $cd_1, cd_2, cd_3 \in Id_{Cd}$.

The relation

$$TraceCd(SourceLayer, SourceCondition, TargetLayer, \\ TargetCondition) \subseteq Id_{La} \times Id_{Cd} \times Id_{La} \times Id_{Cd} \tag{4.82}$$

specifies which trace records for conditions were generated.

The notation

$$cd_1 \xrightarrow[la_1 \to la_2]{Cd} cd_2 :\Leftrightarrow (la_1, cd_1, la_2, cd_2) \in TraceCd \tag{4.83}$$

allows to specify trace records for conditions in an alternative way.

For the condition trace, the following must hold:

$$cd_1 \xrightarrow[la_1 \to la_2]{Cd} cd_2 \wedge cd_1 \xrightarrow[la_1 \to la_2]{Cd} cd_3 \Rightarrow cd_2 = cd_3 \tag{4.84}$$

$$cd_1 \xrightarrow[la_1 \to la_2]{Cd} cd_3 \in TraceCd \wedge cd_2 \xrightarrow[la_1 \to la_2]{Cd} cd_3 \Rightarrow cd_1 = cd_2 \tag{4.85}$$

$$cd_1 \xrightarrow[la_1 \to la_2]{Cd} cd_2 \Rightarrow la_1 = la_2 - 1 \tag{4.86}$$

$$cd_1 \xrightarrow[la_1 \to la_2]{Cd} cd_2 \Rightarrow cd_1 \in conditionsOfLayer(la_1) \tag{4.87}$$
$$\wedge\, cd_2 \in conditionsOfLayer(la_2)$$

Definition 46 (*Action trace*)

Let Id_{La} be the set of layer identifiers, Id_{Ac} the set of action identifiers. Let $la_1, la_2 \in Id_{La}, ac_1, ac_2, ac_3 \in Id_{Ac}$.

The relation

$$TraceAc(SourceLayer, SourceAction, TargetLayer,$$
$$TargetAction) \subseteq Id_{La} \times Id_{Ac} \times Id_{La} \times Id_{Ac} \tag{4.88}$$

specifies which trace records for actions were generated.

The notation

$$ac_1 \xrightarrow[la_1 \to la_2]{Ac} ac_2 :\Leftrightarrow (la_1, ac_1, la_2, ac_2) \in TraceAc \tag{4.89}$$

allows to specify trace records for actions in an alternative way.

For the action trace, the following must hold:

$$ac_1 \xrightarrow[la_1 \to la_2]{Ac} ac_2 \in TraceAc \wedge ac_1 \xrightarrow[la_1 \to la_2]{Ac} ac_3 \Rightarrow ac_2 = ac_3 \tag{4.90}$$

$$ac_1 \xrightarrow[la_1 \to la_2]{Ac} ac_3 \in TraceAc \wedge ac_2 \xrightarrow[la_1 \to la_2]{Ac} ac_3 \Rightarrow ac_1 = ac_2 \tag{4.91}$$

$$ac_1 \xrightarrow[la_1 \to la_2]{Ac} ac_2 \Rightarrow la_1 = la_2 - 1 \tag{4.92}$$

$$ac_1 \xrightarrow[la_1 \to la_2]{Ac} ac_2 \Rightarrow ac_1 \in actionsOfLayer(la_1) \tag{4.93}$$
$$\wedge\, ac_2 \in actionsOfLayer(la_2)$$

Definition 47 (*Operation invoking trace*)

Let Id_{La} be the set of layer identifiers, Id_{OI} the set of operation invoking identifiers. Let $la_1, la_2 \in Id_{La}$, $oi_1, oi_2, oi_3 \in Id_{OI}$.

The relation

$$TraceOI(SourceLayer, SourceOpInv, TargetLayer,$$
$$TargetOpInv) \subseteq Id_{La} \times Id_{OI} \times Id_{La} \times Id_{OI} \tag{4.94}$$

specifies which trace records for operation invokings were generated.

The notation

$$oi_1 \xrightarrow[la_1 \to la_2]{OI} oi_2 :\Leftrightarrow (la_1, oi_1, la_2, oi_2) \in TraceOI \tag{4.95}$$

allows to specify trace records for operation invokings in an alternative way.

For the operation invoking trace, the following must hold:

$$oi_1 \xrightarrow[la_1 \to la_2]{OI} oi_2 \land oi_1 \xrightarrow[la_1 \to la_2]{OI} oi_3 \Rightarrow oi_2 = oi_3 \tag{4.96}$$

$$oi_1 \xrightarrow[la_1 \to la_2]{OI} oi_3 \land oi_2 \xrightarrow[la_1 \to la_2]{OI} oi_3 \Rightarrow oi_1 = oi_2 \tag{4.97}$$

$$oi_1 \xrightarrow[la_1 \to la_2]{OI} oi_2 \Rightarrow la_1 = la_2 - 1 \tag{4.98}$$

$$oi_1 \xrightarrow[la_1 \to la_2]{OI} oi_2 \Rightarrow oi_1 \in opInvsOfLayer(la_1)$$
$$\land oi_2 \in opInvsOfLayer(la_2) \tag{4.99}$$

Refinement Rule Types

Refinement rules refine the policy and linking models by generating refined tuples and inserting them into the respective relations. We distinguish the refinement rules into rules for the policy model and rules for the linking model. Furthermore, we distinguish them by means of the relation on which they operate. Any refinement rule for the policy model operates on one out of the relations *Policy*, *Event*, *Condition*, *ConditionNesting*, and *Action* (3.66). Any refinement rule for the linking model operates on one out of the relations *EL*, *BEL*, *OEL*, *AL*, *OIL1*, *OIL2*, and *OIL3* (3.114).

Table 4.5: Refinement rule types

Model	Relation	Definition
	Policy	Definition 48
Policy	*Event*	Definition 49
model	*Condition, ConditionNesting*	Definition 50
	Action	Definition 51
	EL	Definition 52
Linking	*BEL*	Definition 53
model	*OEL*	Definition 54
	AL	Definition 55
	OIL1, OIL2, OIL3	Definition 56

Table 4.5 summarizes the different refinement rule types and refers to the respective definitions. In the following, we formally define the refinement rules with a relational algebra and thus the formal semantics of the refinement process. The starting points for the refinement are the mapping model and the policy and linking models. These models represent the inputs to the LHSs of the refinement rules, and the refined policy and linking models are implied from them by the RHSs of the refinement rules.

Policies

Definition (48) provides the rule for the refinement of policies. (4.100) transforms any policy po_1 (relation $Policy_{la}$) at any layer except the lowest one into a refined policy po_2 (relation $Policy_{la+1}$) at one layer below. The active status of the policy remains the same in the refinement. In its postcondition, the refinement rule records that po_1 was refined into po_2.

Definition 48 (*Refinement rule for policies*)

Let Val_B be the set of boolean values. Let Id_{La} be the set of layer identifiers, Id_{Po} the set of policy identifiers. Let $Policy_{la}, Policy_{la+1}$ be views of a relation of policies. Let $act \in Val_B, la, lowestLayer \in Id_{La}, po_1, po_2 \in Id_{Po}$. Let $lowestLayer$ be the lowest abstraction layer $max(Id_{La})$.

$$\forall la, po_1, act.$$
$$(la, po_1, act) \in Policy_{la} \wedge la < lowestLayer$$
$$\Rightarrow \exists po_2.$$
$$(la + 1, po_2, act) \in Policy_{la+1}$$
$$\wedge po_1 \xrightarrow[la_1 \to la+1]{Po} po_2$$

(4.100)

Events

Definition (49) provides the rule for the refinement of events. (4.101) requires as a precondition that a policy was refined. Then, it transforms any event ev_1 associated with the respective policy (po_1) into a refined event (ev_2) associated with the respective refined policy (po_2). The refinement rule records in its postcondition that ev_1 was refined into ev_2.

Definition 49 (*Refinement rule for events*)

Let Id_{La} be the set of layer identifiers, Id_{Po} the set of policy identifiers, Id_{Ev} the set of event identifiers. Let $Event_{la_1}, Event_{la_2}$ be views of a relation of events. Let $la_1, la_2 \in Id_{La}, po_1, po_2 \in Id_{Po}, ev_1, ev_2 \in Id_{Ev}$.

$$\forall la_1, la_2, po_1, po_2, ev_1.$$

$$po_1 \xrightarrow[la_1 \to la_2]{Po} po_2$$

$$\wedge \, (po_1, ev_1) \in Event_{la_1}$$

$$\Rightarrow \exists ev_2.$$

$$(po_2, ev_2) \in Event_{la_2}$$

$$\wedge \, ev_1 \xrightarrow[la_1 \to la_2]{Ev} ev_2$$

$$(4.101)$$

Conditions

Definition (50) provides the rule for the refinement of conditions. (4.102) requires as a precondition that a policy was refined. Then, it transforms any condition (cd_1) associated with the respective policy (po_1) into a refined condition (cd_2) associated with the respective refined policy (po_2). The type of a condition remains the same in the refinement. In its postcondition, the refinement rule records that cd_1 was refined into cd_2. (4.103) requires as a precondition that a condition was refined. Then, it transforms any condition (cd_1) nested in the respective condition (cd_2) into a refined condition (cd_3) nested in the respective refined condition (cd_4). The type of a nested condition remains the same in the refinement. In its postcondition, the refinement rule records that cd_1 was refined into cd_3.

Definition 50 (*Refinement rule for conditions*)

Let Id_{La} be the set of layer identifiers, Id_{Po} the set of policy identifiers, Id_{Cd} the set of condition identifiers. Let $Type_{Cd}$ be the set of condition types. Let $Condition_{la_1}, Condition_{la_2}$ be views of a relation of conditions. Let $la_1, la_2 \in Id_{La}$, $po_1, po_2 \in Id_{Po}$, $cd_1, cd_2, cd_3, cd_4 \in Id_{Cd}$, $ty \in Type_{Cd}$.

$$\forall la_1, la_2, po_1, po_2, cd_1, ty.$$
$$po_1 \xrightarrow[la_1 \to la_2]{Po} po_2$$
$$\land (po_1, cd_1, ty) \in Condition_{la_1}$$
$$\Rightarrow \exists cd_2.$$
$$(po_2, cd_2, ty) \in Condition_{la_2}$$
$$\land cd_1 \xrightarrow[la_1 \to la_2]{Cd} cd_2$$

(4.102)

$$\forall la_1, la_2, cd_1, cd_2, cd_3, ty, rc_2.$$
$$cd_2 \xrightarrow[la_1 \to la_2]{Cd} cd_4$$
$$\land (cd_2, cd_1) \in ConditionNesting_{la_1}$$
$$\land (undef, cd_1, ty) \in Condition_{la_1}$$
$$\Rightarrow \exists cd_3.$$
$$(cd_4, cd_3) \in ConditionNesting_{la_2}$$
$$\land (undef, cd_3, ty) \in Condition_{la_2}$$
$$\land cd_1 \xrightarrow[la_1 \to la_2]{Cd} cd_3$$

(4.103)

Actions

Definition (51) provides the rule for the refinement of actions. (4.104) requires as a precondition that a policy was refined. Then, it transforms any action (ac_1) associated with the respective policy (po_1) into a refined action (ac_2) associated with the respective refined policy (po_2). The execution number of an action remains the same in the refinement. The refinement rule records in its postcondition that ac_1 was refined into ac_2.

Definition 51 (*Refinement rule for actions*)

Let Id_{La} be the set of layer identifiers, Id_{Po} the set of policy identifiers, Id_{Ac} the set of action identifiers. Let $Action_{la_1}, Action_{la_2}$ be views of a relation of actions. Let $la_1, la_2 \in Id_{La}$, $po_1, po_2 \in Id_{Po}$, $ac_1, ac_2 \in Id_{Ac}$, $no \in \mathbb{N}$.

$$\forall la_1, la_2, po_1, po_2, ac_1, no.$$

$$po_1 \xrightarrow[la_1 \to la_2]{Po} po_2$$

$$\wedge (po_1, ac_1, no) \in Action_{la_1}$$

$$\Rightarrow \exists ac_2.$$

$$(po_2, ac_2, no) \in Action_{la_2}$$

$$\wedge ac_1 \xrightarrow[la_1 \to la_2]{Ac} ac_2$$

(4.104)

Event Links

Definition (52) provides the rule for the refinement of event links. (4.105) requires as a precondition that an event was refined and a concept mapping was specified. Then, it transforms any event link (relation EL_{la_1}) that refers to the respective event (ev_1) and concept (co_1) into a refined event link (relation EL_{la_2}) that refers to the respective refined event (ev_2) and refined concept (co_2). If the identity, replacement, or merge patterns are used for the concept mapping, the refinement is deterministic with respect to the refined concept and creates one refined event link. If the split pattern is used for the concept mapping, multiple possibilities exist for the concept in the refined event link. In this case, the refinement is non-deterministic with respect to the refined concept and specifies a choice of the different possibilities as an exclusive disjunction ($\oplus$). If the erasure pattern is used for the concept mapping, the refinement rule does not apply, i.e. no refined event link is created.

Definition 52 (*Refinement rule for event links*)

Let Id_{La} be the set of layer identifiers, Id_{Co} the set of concept identifiers, Id_{Ev} the set of event identifiers. Let EL_{la_1}, EL_{la_2} be views of a relation of event links. Let $la_1, la_2 \in Id_{La}, co_1, co_{2_1}, ..., co_{2_n} \in Id_{Co}, ev_1, ev_2 \in Id_{Ev}$.

$$\forall la_1, la_2, ev_1, ev_2, co_1, co_{2_1}, ..., co_{2_n}.$$

$$ev_1 \xrightarrow[la_1 \to la_2]{Ev} ev_2 \wedge co_1 \xrightarrow[la_1 \to la_2]{Co} (co_{2_1}, ..., co_{2_n})$$

$$\wedge (ev_1, co_1) \in EL_{la_1}$$

$$\Rightarrow \bigoplus_{i \in \{1,...,n\} \wedge co_{2_i} \neq undef} (ev_2, co_{2_i}) \in EL_{la_2}$$

(4.105)

Binary Expression Links

Definition (53) provides the rule for the refinement of binary expression links. (4.106) to (4.114) require as a precondition that a binary expression was refined. Then, they transform any binary expression link (relation BEL_{la_1}) that refers to the respective binary expression (be_1) and first and second arguments into a refined binary expression link (relation BEL_{la_2}) that refers to the respective refined binary expression (be_2) and refined first and second arguments.

A binary expression link can use a literal, a property, or an operation invoking as first or second argument, and each combination is covered by one case. If both arguments are literals, (4.106) reuses the same literals as refined arguments. If both arguments are properties, (4.107) requires as additional precondition that respective property mappings were specified and uses the refined properties as refined arguments. If both arguments are operation invokings, (4.108) triggers the refinement of these operation invokings and uses the refined invokings as refined arguments.

If one argument is a property and the other argument is a literal, (4.109) and (4.110) require as additional precondition that a respective property mapping was specified. The literal is refined according to the source property, the source literal, and the target property. The refined property and refined literal are used as refined arguments.

If one argument is an operation invoking and the other argument is a literal, (4.111) and (4.112) trigger the refinement of this operation invoking. The literal is refined according to the invoked source operation, the source literal, and the invoked target operation. The refined operation invoking and refined literal are used as refined arguments.

If one argument is an operation invoking and the other argument is a property, (4.113) and (4.114) require as additional precondition that a respective property mapping was specified and trigger the refinement of this operation invoking. The refined operation invoking and property are used as refined arguments.

If the identity, replacement, or merge patterns are used for the property mappings, the refinement is deterministic with respect to the refined properties and creates one refined binary expression link. If the split pattern is used for the property mappings, multiple possibilities exist for the properties in the refined binary expression link. In this case, the refinement is nondeterministic with respect to the refined properties and specifies a choice of the different possibilities as an exclusive disjunction ($\oplus$). If the erasure pattern is used for the property mappings, the refinement rule does not apply, i.e. no refined binary expression link is created. Different possibilities can also apply to the refined literals. If a source literal is mapped to different target literals in the literal mapping, the refinement rule is non-deterministic with respect to the refined literals and specifies a choice of the different possibilities.

Definition 53 (*Refinement rule for binary expression links*)

Let Lit be the set of literals. Let Id_{La} be the set of layer identifiers, Id_{Pr} the set of property identifiers, Id_{Cd} the set of condition identifiers, Id_{OI} the set of operation invoking identifiers. Let BEL_{la_1}, BEL_{la_2} be views of a relation of binary expression links. Let $la_1, la_2 \in Id_{La}$, $pr_1, pr_2, pr_{2_1}, ..., pr_{2_n}, pr_{3_1}, ..., pr_{3_m}, pr_{4_1}, ..., pr_{4_n} \in Id_{Pr}$, $be_1, be_2 \in Id_{Cd}$, $oi_1, oi_2, oi_3, oi_4 \in Id_{OI}$. Let $lit_1, lit_2 \in Lit$.

$$\forall la_1, la_2, be_1, be_2, lit_1, lit_2.$$

$$be_1 \xrightarrow[la_1 \to la_2]{Cd} be_2 \tag{4.106}$$

$$\wedge (be_1, lit_1, lit_2) \in BEL_{la_1} \wedge lit_1 \in Lit \wedge lit_2 \in Lit$$

$$\Rightarrow (be_2, lit_1, lit_2) \in BEL_{la_2}$$

$$\forall la_1, la_2, be_1, be_2, pr_1, pr_2, pr_{3_1}, ..., pr_{3_m}, pr_{4_1}, ..., pr_{4_n}.$$

$$be_1 \xrightarrow[la_1 \to la_2]{Cd} be_2 \wedge pr_1 \xrightarrow[la_1 \to la_2]{Pr} (pr_{3_1}, ..., pr_{3_m})$$

$$\wedge pr_2 \xrightarrow[la_1 \to la_2]{Pr} (pr_{4_1}, ..., pr_{4_n}) \tag{4.107}$$

$$\wedge (be_1, pr_1, pr_2) \in BEL_{la_1} \wedge pr_1 \in Id_{Pr} \wedge pr_2 \in Id_{Pr}$$

$$\Rightarrow \bigoplus_{\substack{i \in \{1,...,m\} \wedge pr_{3_i} \neq undef \\ \wedge j \in \{1,...,n\} \wedge pr_{4_j} \neq undef}} (be_2, pr_{3_i}, pr_{4_j}) \in BEL_{la_2}$$

$$\forall la_1, la_2, be_1, be_2, oi_1, oi_2.$$

$$be_1 \xrightarrow[la_1 \to la_2]{Cd} be_2$$

$$\wedge (be_1, oi_1, oi_2) \in BEL_{la_1} \wedge oi_1 \in Id_{OI} \wedge oi_2 \in Id_{OI} \tag{4.108}$$

$$\Rightarrow \exists oi_3, oi_4.$$

$$(be_2, oi_3, oi_4) \in BEL_{la_2}$$

$$\wedge oi_1 \xrightarrow[la_1 \to la_2]{OI} oi_3 \wedge oi_2 \xrightarrow[la_1 \to la_2]{OI} oi_4$$

$\forall la_1, la_2, be_1, be_2, pr_1, pr_{2_1}, ..., pr_{2_n}, lit_1.$

$\quad be_1 \xrightarrow[la_1 \to la_2]{Cd} be_2 \wedge pr_1 \xrightarrow[la_1 \to la_2]{Pr} (pr_{2_1}, ..., pr_{2_n})$

$\quad \wedge (be_1, pr_1, lit_1) \in BEL_{la_1} \wedge pr_1 \in Id_{Pr} \wedge lit_1 \in Lit$

$\quad \Rightarrow \exists lit_2.$

$$\bigoplus_{i \in \{1, ..., n\} \wedge pr_{2_i} \neq undef} (be_2, pr_{2_i}, lit_2) \in BEL_{la_2}$$

$\quad\quad \wedge lit_2 \in mappedLiterals^{Pr}_{la_1 \to la_2}(pr_1, lit_1, pr_{2_i})$
$\hfill (4.109)$

$\forall la_1, la_2, be_1, be_2, lit_1, pr_1, pr_{2_1}, ..., pr_{2_n}.$

$\quad be_1 \xrightarrow[la_1 \to la_2]{Cd} be_2 \wedge pr_1 \xrightarrow[la_1 \to la_2]{Pr} (pr_{2_1}, ..., pr_{2_n})$

$\quad \wedge (be_1, lit_1, pr_1) \in BEL_{la_1} \wedge lit_1 \in Lit \wedge pr_1 \in Id_{Pr}$

$\quad \Rightarrow \exists lit_2.$

$$\bigoplus_{i \in \{1, ..., n\} \wedge pr_{2_i} \neq undef} (be_2, lit_2, pr_{2_i}) \in BEL_{la_2}$$

$\quad\quad \wedge lit_2 \in mappedLiterals^{Pr}_{la_1 \to la_2}(pr_1, lit_1, pr_{2_i})$
$\hfill (4.110)$

$\forall la_1, la_2, be_1, be_2, oi_1, lit_1.$

$\quad be_1 \xrightarrow[la_1 \to la_2]{Cd} be_2$

$\quad \wedge (be_1, oi_1, lit_1) \in BEL_{la_1} \wedge oi_1 \in Id_{OI} \wedge lit_1 \in Lit$

$\quad \Rightarrow \exists oi_2, lit_2.$

$\quad\quad (be_2, oi_2, lit_2) \in BEL_{la_2} \wedge lit_2 \in mappedLiterals^{Op}_{la_1 \to la_2}$

$\quad\quad\quad (operationOfOpInv(oi_1), lit_1, operationOfOpInv(oi_2))$

$\quad\quad \wedge oi_1 \xrightarrow[la_1 \to la_2]{OI} oi_2$
$\hfill (4.111)$

$\forall la_1, la_2, be_1, be_2, lit_1, oi_1.$

$\quad be_1 \xrightarrow[la_1 \to la_2]{Cd} be_2$

$\quad \wedge (be_1, lit_1, oi_1) \in BEL_{la_1} \wedge lit_1 \in Lit \wedge oi_1 \in Id_{OI}$

$\quad \Rightarrow \exists lit_2, oi_2.$

$\quad\quad (be_2, lit_2, oi_2) \in BEL_{la_2} \wedge lit_2 \in mappedLiterals^{Op}_{la_1 \to la_2}$

$\quad\quad\quad (operationOfOpInv(oi_1), lit_1, operationOfOpInv(oi_2))$

$\quad\quad \wedge oi_1 \xrightarrow[la_1 \to la_2]{OI} oi_2$
$\hfill (4.112)$

$$\forall la_1, la_2, be_1, be_2, oi_1, pr_1, pr_{2_1}, ..., pr_{2_n}.$$

$$be_1 \xrightarrow[la_1 \to la_2]{Cd} be_2 \wedge pr_1 \xrightarrow[la_1 \to la_2]{Pr} (pr_{2_1}, ..., pr_{2_n})$$

$$\wedge (be_1, oi_1, pr_1) \in BEL_{la_1} \wedge oi_1 \in Id_{OI} \wedge pr_1 \in Id_{Pr}$$

$$\Rightarrow \exists oi_2. \tag{4.113}$$

$$\bigoplus_{i \in \{1,...,n\} \wedge pr_{2_i} \neq undef} (be_2, oi_2, pr_{2_i}) \in BEL_{la_2}$$

$$\wedge oi_1 \xrightarrow[la_1 \to la_2]{OI} oi_2$$

$$\forall la_1, la_2, be_1, be_2, pr_1, pr_{2_1}, ..., pr_{2_n}, oi_1.$$

$$be_1 \xrightarrow[la_1 \to la_2]{Cd} be_2 \wedge pr_1 \xrightarrow[la_1 \to la_2]{Pr} (pr_{2_1}, ..., pr_{2_n})$$

$$\wedge (be_1, pr_1, oi_1) \in BEL_{la_1} \wedge pr_1 \in Id_{Pr} \wedge oi_1 \in Id_{OI}$$

$$\Rightarrow \exists oi_2. \tag{4.114}$$

$$\bigoplus_{i \in \{1,...,n\} \wedge pr_{2_i} \neq undef} (be_2, pr_{2_i}, oi_2) \in BEL_{la_2}$$

$$\wedge oi_1 \xrightarrow[la_1 \to la_2]{OI} oi_2$$

Operational Expression Links

Definition (54) provides the rule for the refinement of operational expression links. (4.115) requires as a precondition that an operational expression was refined. Then, it transforms any operational expression link (relation OEL_{la_1}) that refers to the respective operational expression (oe_1) and an operation invoking (oi_1) into a refined operational expression link (relation OEL_{la_2}) that refers to the respective refined operational expression (oe_2) and a refined operation invoking (oi_2). In its postcondition, the refinement rule records that oi_1 was refined into oi_2.

Definition 54 (*Refinement rule for operational expression links*)

Let Id_{La} be the set of layer identifiers, Id_{Cd} the set of condition identifiers, Id_{OI} the set of operation invoking identifiers. Let OEL_{la_1}, OEL_{la_2} be views of a relation of operational expression links. Let $la_1, la_2 \in Id_{La}$, $oi_1, oi_2 \in Id_{OI}$, $oe_1, oe_2 \in Id_{Cd}$.

$$\forall la_1, la_2, oe_1, oe_2, oi_1.$$

$$oe_1 \xrightarrow[la_1 \to la_2]{Cd} oe_2$$

$$\wedge (oe_1, oi_1) \in OEL_{la_1}$$

$$\Rightarrow \exists oi_2.$$

$$(oe_2, oi_2) \in OEL_{la_2}$$

$$\wedge oi_1 \xrightarrow[la_1 \to la_2]{OI} oi_2$$

(4.115)

Action Links

Definition (55) provides the rule for the refinement of operational expression links. (4.116) requires as a precondition that an action was refined. Then, it transforms any action link (relation AL_{la_1}) that refers to the respective action (ac_1) and an operation invoking (oi_1) into a refined action link (relation AL_{la_2}) that refers to the respective refined action (ac_2) and a refined operation invoking (oi_2). The refinement rule records in its postcondition that oi_1 was refined into oi_2.

Definition 55 (*Refinement rule for action links*)

Let Id_{La} be the set of layer identifiers, Id_{Ac} the set of action identifiers, Id_{OI} the set of operation invoking identifiers. Let $OIL1$ be a first relation of operation invokings. Let $la_1, la_2 \in Id_{La}$, $ac_1, ac_2 \in Id_{Ac}$, $oi_1, oi_2 \in Id_{OI}$.

$$\forall la_1, la_2, ac_1, ac_2, oi_1.$$

$$ac_1 \xrightarrow[la_1 \to la_2]{Ac} ac_2$$

$$\wedge (ac_1, oi_1) \in AL_{la_1}$$

$$\Rightarrow \exists oi_2.$$

$$(ac_2, oi_2) \in AL_{la_2}$$

$$\wedge oi_1 \xrightarrow[la_1 \to la_2]{OI} oi_2$$

(4.116)

Operation Invoking Links

Definition (56) provides the rule for the refinement of operational invokings. An operation invoking link consists of an operation invoking and links to an operation, parameters, and arguments.

(4.117) defines the refinement of the operation invoking link. It requires as a precondition that an operation invoking was refined. Then, it transforms any operation invoking link (relation $OIL1_{la_1}$) that refers to the respective operation invoking (oi_1) into a refined operation invoking link (relation $OIL2_{la_2}$) that refers to a refined operation invoking (oi_2).

(4.118) defines the refinement of the linked operation. It requires as a precondition that an operation invoking was refined and an operation mapping was specified. Then, it transforms any operation invoking link (relation $OIL2_{la_1}$) that refers to the respective operation invoking (oi_1) and operation (op_1) into a refined operation invoking link (relation $OIL2_{la_1}$) that refers to the respective refined operation invoking (oi_2) and refined operation (op_2). If the identity, replacement, or merge patterns are used for the operation mapping, the refinement is deterministic with respect to the refined operation and creates one refined operation invoking link. If the split pattern is used for the operation mapping, multiple possibilities exist for the operation in the refined operation invoking link. In this case, the refinement is non-deterministic with respect to the refined operation and specifies a choice of the different possibilities as an exclusive disjunction ($\oplus$). If the erasure pattern is used for the operation mapping, the refinement rule does not apply, i.e. no refined operation invoking link is created.

(4.119) to (4.121) define the refinement of the linked parameters and linked arguments. If the argument is a literal, (4.119) refines the literal according to the source parameter, the source literal, and the target parameter. The refined literal is used as refined argument. If the argument is a property, (4.120) requires as an additional precondition that a property mapping was specified and uses the refined property as refined argument. If the argument is an operation invoking, (4.121) triggers the refinement of this operation invoking and uses the refined invoking as refined argument. In any case, the refinement rule requires as a precondition that an operation invoking was refined and a parameter mapping was specified. Then, it transforms any operation invoking link (relation $OIL3_{la_1}$) that refers to the respective operation invoking (oi_1), parameter (pa_1), and argument (lit, pr_1, or oi_2) into a refined operation invoking link (relation $OIL2_{la_1}$) that refers to the respective refined operation invoking (oi_2), refined parameter (pa_2), and refined argument (lit, pr_2, or oi_4). The refinement rule states as a postcondition that oi_2 was refined into oi_4. If the identity, replacement, or merge patterns are used for the parameter mapping, the refinement is deterministic with respect to the refined parameter and create one refined operation invoking link. If the split pattern is used for the parameter mapping, multiple possibilities exist for the parameter in the refined operation invoking link. In this case, the refinement is non-deterministic with respect to the refined parameter and specifies a choice of the different possibilities as an exclusive disjunction ($\oplus$). If the erasure pattern is used for the parameter mapping, the refinement rule does not apply, i.e. no refined operation invoking link is created.

(4.122) defines the creation of new parameters and arguments with refined operation links. It requires as precondition that an operation invoking was refined and a parameter appearance was specified with the operation of the refined operation invoking link. Then, it creates a new operation invoking link (relation $OIL3_{la_1}$) that refers to the respective refined operation invoking (oi_2), refined parameter (pa_2), and a refined literal (lit_2). The literal is chosen according to the appeared literal in the respective literal mapping. Different possibilities can apply for the appeared literal. In this case, the refinement rule is non-deterministic with respect to the appeared literal and again specifies a choice of the different possibilities.

Definition 56 (*Refinement rule for operation invoking links*)

Let Id_{La} be the set of layer identifiers, Id_{Pr} the set of property identifiers. Id_{Op} the set of operation identifiers, Id_{Pa} the set of parameter identifiers, Id_{OI} the set of operation invoking identifiers. Let $OIL1_{la_1}, OIL1_{la_2}, OIL2_{la_1}, OIL2_{la_2}, OIL3_{la_1}, OIL3_{la_2}$ be views of relations of operation invoking links. Let $lit_1, lit_2 \in Lit$, $la_1, la_2 \in Id_{La}$, $pr_1, pr_{2_1}, ..., pr_{2_n} \in Id_{Pr}$, $op_1, op_{2_1}, ..., op_{2_n} \in Id_{Op}$, $pa_1, pa_{2_1}, ..., pa_{2_m}, pa_{2_1}, ..., pa_{2_n}, pa_{3_1}, ..., pa_{3_n} \in Id_{Pa}$, $oi_1, oi_2, oi_3, oi_4 \in Id_{OI}$.

$\forall la_1, la_2, oi_1, oi_2.$

$$oi_1 \xrightarrow[la_1 \to la_2]{OI} oi_2$$
$$\land oi_1 \in OIL1_{la_1}$$
$$\Rightarrow oi_2 \in OIL1_{la_2}$$

(4.117)

$\forall la_1, la_2, oi_1, oi_2, op_1, op_{2_1}, ..., op_{2_n}.$

$$oi_1 \xrightarrow[la_1 \to la_2]{OI} oi_2 \land op_1 \xrightarrow[la_1 \to la_2]{Op} (op_{2_1}, ..., op_{2_n})$$
$$\land (oi_1, op_1) \in OIL2_{la_1}$$
$$\Rightarrow \bigoplus_{i \in \{1,...,n\} \land op_{2_i} \neq undef} (oi_2, op_{2_i}) \in OIL2_{la_2}$$

(4.118)

$$\forall la_1, la_2, oi_1, oi_2, pa_1, pa_{2_1}, ..., pa_{2_n}, lit_1.$$

$$oi_1 \xrightarrow[la_1 \to la_2]{OI} oi_2 \wedge pa_1 \xrightarrow[la_1 \to la_2]{Pa} (pa_{2_1}, ..., pa_{2_n})$$

$$\wedge (oi_1, pa_1, lit_1) \in OIL3_{la_1} \wedge lit_1 \in Lit$$

$$\Rightarrow \exists lit_2. \tag{4.119}$$

$$\bigoplus_{\substack{i \in \{1,...,n\} \wedge pa_{2_i} \neq undef \wedge (oi_2, op_2) \in OIL2_{la_1} \\ \wedge op_2 = operationOfParameter(pa_{2_i})}} (oi_2, pa_{2_i}, lit_2) \in OIL3_{la_2}$$

$$\wedge lit_2 \in mappedLiterals^{Pa}_{la_1 \to la_2}(pa_1, lit_1, pa_{2_i})$$

$$\forall la_1, la_2, oi_1, oi_2, pa_1, pa_{3_1}, ..., pa_{3_n}, pr_1, pr_{2_1}, ..., pr_{2_n}.$$

$$oi_1 \xrightarrow[la_1 \to la_2]{OI} oi_2 \wedge pa_1 \xrightarrow[la_1 \to la_2]{Pa} (pa_{2_1}, ..., pa_{2_m})$$

$$\wedge pr_1 \xrightarrow[la_1 \to la_2]{Pr} (pr_{2_1}, ..., pr_{2_n})$$

$$\wedge (oi_1, pa_1, pr_1) \in OIL3_{la_1} \wedge pr_1 \in Id_{Pr} \tag{4.120}$$

$$\Rightarrow \bigoplus_{\substack{i \in \{1,...,m\} \wedge pa_{2_i} \neq undef \wedge (oi_2, op_2) \in OIL2_{la_1} \\ \wedge op_2 = operationOfParameter(pa_{2_i}) \\ \wedge j \in \{1,...,n | pr_{2_j} \neq undef\}}} (oi_2, pa_{2_i}, pr_{2_j}) \in OIL3_{la_2}$$

$$\forall la_1, la_2, oi_1, oi_2, oi_3, pa_1, pa_{3_1}, ..., pa_{3_n}.$$

$$oi_1 \xrightarrow[la_1 \to la_2]{OI} oi_3 \wedge pa_1 \xrightarrow[la_1 \to la_2]{Pa} (pa_{3_1}, ..., pa_{3_m})$$

$$\wedge (oi_1, pa_1, oi_2) \in OIL3_{la_1} \wedge oi_2 \in Id_{OI}$$

$$\Rightarrow \exists oi_4. \tag{4.121}$$

$$\bigoplus_{\substack{i \in \{1,...,n\} \wedge pa_{3_i} \neq undef \wedge (oi_3, op_3) \in OIL2_{la_1} \\ \wedge op_3 = operationOfParameter(pa_{3_i})}} (oi_3, pa_{3_i}, oi_4) \in OIL3_{la_2}$$

$$\wedge oi_2 \xrightarrow[la_1 \to la_2]{OI} oi_4$$

$$\forall la_1, la_2, oi_1, oi_2, pa_{2_1}, ..., pa_{2_n}, lit_1.$$

$$oi_1 \xrightarrow[la_1 \to la_2]{OI} oi_2 \wedge \xrightarrow[la_1 \to la_2]{Pa} (pa_{2_1}, ..., pa_{2_n}) \wedge \forall i \in \{1, ..., n\}.$$

$$pa_{2_i} \in parametersOfOperation(operationOfOpInv(oi_2))$$

$$\Rightarrow \exists lit_2. \tag{4.122}$$

$$\bigoplus_{\substack{i \in \{1,...,n\} \wedge pa_{2_i} \neq undef \wedge (oi_2, op_2) \in OIL2_{la_1} \\ \wedge op_2 = operationOfParameter(pa_{2_i})}} (oi_2, pa_{2_i}, lit_2) \in OIL3_{la_2}$$

$$\wedge lit_2 \in appearedLiterals^{Pa}_{la_1 \to la_2}(pa_{2_i})$$

4.2.2 Generation of Refined Policies

As soon as the mapping model and the policy and linking models are provided at the highest layer, the refined policy and linking models at the lower layers are implied by the refinement rules. We provided the execution semantics of the refinement implicitly with the declarative refinement rules. Now, we move the point of view from a declarative one to an imperative one and thus provide the explicit execution semantics of the refinement. This is a non-trivial task [41], but helps to better understand the refinement process. Furthermore, imperative execution semantics is useful for an implementation with an imperative model transformation language. The execution semantics of the refinement process from the imperative point of view is illustrated by algorithm 1. This algorithm can be used as a starting point for an imperative implementation of the refinement process. The definition of trivial procedures is omitted in the algorithm. The non-trivial procedures are explained in the following.

- **procedure** REFINEMODEL

  In the beginning, the algorithm determines which layer of the model is the highest one and which is the lowest one. It then iterates over the layers from top to bottom. In each iteration, it refines the view of the policy and linking models at the respective layer into the view at the subsequent lower layer as intermediate result. The algorithm stops after the view at the second lowest layer has been refined, resulting in the view at the lowest layer.

- **procedure** REFINELAYER(*layer*)

  To refine the view at a layer into the view at the subsequent lower layer, the algorithm determines the refinement rules and then applies them to the respective layer one after another. We discuss below that a particular sequence is necessary in this iteration.

- **procedure** APPLYRULE(*rule, layer*)

  Now, a refinement rule is applied to a layer. Preceding refinement rules generated triggers for the current rule, and these triggers are initially determined from the trace. The refinement rule for policies represents an exception as it is not preceded by any rule. In this case, triggers for all policies at the respective layer are returned. The refinement rule then iterates over the triggers, and selects the tuples to be refined for each trigger. It then iteratively refines these tuples and generates additional triggers for other rules. The refined tuples are added to the model, and generated triggers are added to the trace. We discuss below that the sequence of iteration is not relevant for triggers and tuples.

Algorithm 1 Generation of refined policies

---

procedure REFINEMODEL
    highestLayer ← GETHIGHESTLAYER()
    lowestLayer ← GETLOWESTLAYER()
    for *layer* ← *highestLayer* to *lowestLayer* + 1 **do**
        REFINELAYER(*layer*)
    end for
end procedure

procedure REFINELAYER(*layer*)
    rules ← GETRULES()
    for all *rule* ← *rules* **do**
        APPLYRULE(*rule, layer*)
    end for
end procedure

procedure APPLYRULE(*rule, layer*)
    triggers ← GETTRIGGERS(*rule, layer*)
    for all *trigger* ← *triggers* **do**
        tuples ← GETTUPLES(*rule, trigger*)
        for all *tuple* ← *tuples* **do**
            refinedTuple ← REFINETUPLE(*tuple, rule*)
            ADDTUPLE(*refinedTuple*)
            newTrigger ← GENERATETRIGGER(*trigger, rule*)
            ADDTRIGGER(*refinedTrigger, triggers*)
        end for
    end for
end procedure

procedure REFINETUPLE(*tuple, rule*)
    refinedTuples ← GENERATEREFINEDTUPLES(*rule, tuple*)
    if SIZE(*refinedTuples*) = 1 **then**
        refinedTuple ← FIRST(*refinedTuples*)
    else
        refinedTuple ← DECIDE(*refinedTuples*)
    end if
    return *refinedTuple*
end procedure

- **procedure** REFINETUPLE(*tuple, rule*)

  To refine a tuple with a refinement rule, the refined tuples are generated according to the LHS and RHS of the rule. If the respective rule is deterministic, one refined tuple is generated. Otherwise, a set of refined tuples are proposed and an implementation must decide for one out these possibilities, e.g. by offering them to the developer and requesting a decision from him. This feature takes account of cases where a definite refinement is not possible initially, and it allows to delay the decision until the refinement is actually performed.

Sequence of Operation

From this imperative point of view, the procedures of the algorithm perform several iterations. Iterations are performed over layers, refinement rules, triggers, and tuples. As described in the following, the sequence is relevant for some of these iterations, and for some it is not.

- **procedure** REFINEMODEL

  This procedure iterates over the layers for purposes of refinement. The sequence of iteration is crucial here. At first, the view of the policy and linking models at the highest layer is refined into the view at the second highest layer. The view at the second highest layer is further refined by the following iteration, and the algorithm further iterates until the view at the lowest layer is generated. This sequence is crucial as the output of an iteration represents the input for the following iteration.

- **procedure** REFINELAYER(*layer*)

  In this procedure, an iteration is performed over the rules to apply the rules one after another. The sequence of this iteration is crucial as the refinement rules trigger each other. Only with an appropriate sequence, the refined tuples are generated as defined by the formal semantics of the refinement process. Figure 4.9 shows the respective dependencies between the refinement rules. The refinement rule for policies is always applied at first as it is the only one without a trigger, i.e. without a precondition in its formal definition. All other rules can only be applied after all other preceding rules have been applied completely. According to the dependencies, different sequences of application are possible, and any of these leads to the same result. Refinement rules can also be applied in parallel if they do not depend on each other. This enables a certain degree of parallelization in an imperative implementation.

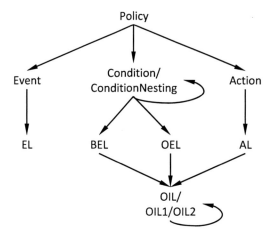

Figure 4.9: Possible sequences for the refinement of the relations

- **procedure** APPLYRULE(*rule, layer*)

  A refinement rule is often triggered multiple times by its preceding
  rules. For this reason, an iteration over multiple triggers is performed
  in this procedure. The purpose of the triggers is to ensure the refinement of all tuples that need to be refined. Most of the rules generate
  additional triggers in their postcondition for other rules or even for
  the rule itself. However, generated triggers do not have any influence
  on the current application of the rule. For this reason, the sequence
  of iteration over the triggers is arbitrary, even if a rule generates additional triggers for itself. The sequence of iteration over the tuples is
  also arbitrary in this procedure as refined tuples are added to another
  layer of the model and do not have any influence on the application
  of the rule on the current tuple. Thus, any sequence of triggers and
  tuples leads to the same result. An imperative implementation could
  even process all triggers and tuples within a rule in parallel.

- **procedure** REFINETUPLE(*tuple, rule*)

  This procedure does not perform any iteration.

Complexity

We now determine the time complexity of the algorithm in big O notation. For this purpose, we first determine the complexity of the innermost procedure and then subsequently determine the complexity of the outer procedures until the complexity of the overall algorithm is determined.

Let *layers* be the number of layers in the domain model, $tuples_{rule,layer}$ be the number of tuples processed by a particular refinement rule at a particular layer, and $tuples_{layer}$ be the overall number of tuples at a particular layer.

- **procedure** REFINETUPLE($tuple, rule$)

  This procedure refines one tuple with one rule and requests a decision if multiple refined links are possible. We assume that the used procedures GENERATEREFINEDTUPLES, SIZE, FIRST, and DECIDE are performed in constant time ($O(1)$). Thus, the time complexity is:

  $$T_{\text{REFINETUPLE}} \in O(1) \tag{4.123}$$

- **procedure** APPLYRULE($rule, layer$)

  In this procedure, a refinement rule is applied to the view of the policy and linking models at a particular layer. We assume that the used procedures GETTRIGGERS, GETTUPLES, and ADDTRIGGER are performed in constant time ($O(1)$). This is possible with appropriate data structures. The procedure processes the tuples according to the respective triggers. An appropriate implementation should recognize multiple triggers for the same tuple and avoid any repeated processing. For a pessimistic calculation, we assume worst case policy and linking models that require the refinement of all tuples contained at the highest layer. Then, the procedure REFINETUPLE ($O(1)$) is invoked for each tuple matched by the refinement rule ($O(tuples_{rule,layer})$). The refined tuples are added to the model, and additional triggers might be generated and added to the trace. We assume that the used procedures ADDTUPLE, GENERATETRIGGER, and ADDTRIGGER are performed in constant time ($O(1)$). Thus, the time complexity is:

  $$T_{\text{APPLYRULE}} \in O(tuples_{rule,layer}) \tag{4.124}$$

An upper bound can be derived for the number of tuples processed by a rule at a layer. The number of tuples is not higher than the overall number of tuples at that layer (4.125). Furthermore, the number of tuples at a layer is not higher than the number of tuples at the highest layer (4.126). This is because a refinement rule never creates more tuples at the target layer than tuples that were available at the source layer (definitions 49 to 56).

$$\forall rule, layer.tuples_{rule,layer} \leq tuples_{layer} \tag{4.125}$$

$$\forall layer.tuples_{layer} \leq tuples_{highestLayer} \tag{4.126}$$

Therefore, the following holds:

$$T_{\text{APPLYRULE}} \in O(tuples_{rule,layer})$$
$$\xRightarrow{(4.125)} T_{\text{APPLYRULE}} \in O(tuples_{layer}) \tag{4.127}$$
$$\xRightarrow{(4.126)} T_{\text{APPLYRULE}} \in O(tuples_{highestLayer})$$

- **procedure** REFINELAYER($layer$)

  This procedure performs the refinement of all rules one after another. For each rule, ($O(1)$) the procedure APPLYRULE ($O(tuples_{highestLayer})$) is invoked. We assume that the used procedure TYPE is performed in constant time ($O(1)$). Thus, the time complexity is:

  $$T_{\text{REFINELAYER}}(tuples_{highestLayer})$$
  $$\subseteq O(tuples_{highestLayer}) \tag{4.128}$$

- **procedure** REFINEMODEL

  In this procedure, the procedure REFINELAYER ($O(tuples_{highestLayer})$) is invoked for each layer except the lowest one ($O(layers)$). We assume that the used procedures GETHIGHESTLAYER and GETLOWESTLAYER are performed in constant time ($O(1)$). Thus, the time complexity is:

  $$T_{\text{REFINEMODEL}}$$
  $$\in O(layers \cdot tuples_{highestLayer}) \tag{4.129}$$

As a result, we showed that the algorithm runs in linear time depending on two influencing factors. These factors are the number of abstraction layers and the number of tuples in the view at the highest layer, as illustrated in figure 4.10. Notably, the time complexity does not depend on the mapping model. This comes from the fact that the mapping model and the refinement rules are designed in a way that even with the most complex mappings, each object is only refined once.

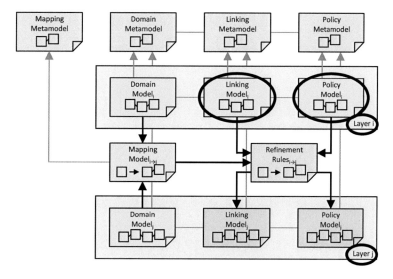

Figure 4.10: Influencing factors for $T_{\text{REFINEMODEL}}$

4.2.3 Validation of the Refinement

As mentioned before, executable code should be generated from the refined policies. For a proper code generation, the validity of the refined models is an essential prerequisite. If the refined policy and linking models are no valid instances of the respective metamodels, no proper code can be generated.

We now show that the refinement rules generate valid models only, in case valid models are provided as input. For this purpose, we perform an induction over the number of abstraction layers and over the possible input of the refinement rules. We then show for all cases that the rules generate a valid model as output again. Thus, we prove that the refinement is syntactically and semantically correct according to definition 57 [170, 171].

Definition 57 (*Correctness*)

> A model transformation is syntactically correct if and exactly if the target model is well-formed, i.e. conforms to the target metamodel. For the policy model, we check syntactical correctness according to (3.68), (3.75), (3.84), (3.95), and (3.105). For the linking model, we check syntactical correctness according to (3.133), (3.142), (3.155), (3.162), and (3.116)–(3.119).
>
> A model transformation is semantically correct if and exactly if the transformation preserves the static semantics of the source model, i.e. the semantic properties also hold for the target model. For the policy model, we check semantical correctness according to the properties (3.69), (3.76), (3.85)–(3.86), (3.96)–(3.101), and (3.106)–(3.107). For the linking model, we check semantical correctness according to the properties (3.134)–(3.136), (3.143)–(3.147), (3.156)–(3.159), (3.163)–(3.166), and (3.120)–(3.123).

In our proof, we firstly perform a first (complete) induction over the number of layers. We start with a refinement over zero layers (base clause 1) and then increase the number of layers by one (induction step 1). In this induction step, we refine the views of the policy and linking models at layer la into the views at the subsequent layer $la + 1$. To capture all possible views of policy and linking models, we perform a second (structural) induction over the views, i.e. we start with empty views (base clause 2) and then extend them in any possible way (induction step 2). For each possible extension, we apply the refinement rules and show that the syntactical and semantical correctness is preserved.

Statement

Let DM be a domain model, MM a mapping model, PM a policy model, LM a linking model.

Then, the following holds:

DM is valid $\wedge$ MM is valid

$\wedge$ $PM_{highestLayer}$ is a valid view of PM (cf. definition 9)

    and refined into $PM_{lowestLayer}$ by (4.100)–(4.122)

$\wedge$ $LM_{highestLayer}$ is a valid view of LM (cf. definition 15)   (4.130)

    and refined into $LM_{lowestLayer}$ by (4.100)–(4.122)

$\Rightarrow$ $PM_{lowestLayer}$ is a valid view of PM (cf. definition 9)

    $\wedge$ $LM_{lowestLayer}$ is a valid view of LM (cf. definition 15)

Base Clause 1

Let the hightest and lowest layer be the same, i.e. $highestLayer = lowestLayer$. If $PM_{highestLayer}$ and $LM_{highestLayer}$ are valid views of PM and LM, then $PM_{lowestLayer}$ and $LM_{lowestLayer}$ are still valid views of PM and LM.

Induction Hypothesis 1

Let (4.130) hold for particular la, DM, MM, $PM_{highestLayer}$, $LM_{highestLayer}$, PM_{la}, and LM_{la}, i.e. valid views $PM_{highestLayer}$ and $LM_{highestLayer}$ were refined by (4.100)–(4.122) into valid views PM_{la} and LM_{la}.

Induction Step 1

We now increase the number of layers by one, i.e. we add a layer $la + 1$. We then refine PM_{la} and LM_{la} by (4.100)–(4.122) into PM_{la+1} and LM_{la+1}. Finally, we show that PM_{la+1} and LM_{la+1} are still valid views of PM and LM, i.e. that (4.130) still holds.

Base Clause 2

Let PM and LM be empty. Then, PM_{la} and LM_{la} are valid views of PM and LM:

$$PM_{la} = (\emptyset, \emptyset, \emptyset, \emptyset, \emptyset)$$
$$LM_{la} = (\emptyset, \emptyset, \emptyset, \emptyset, \emptyset, \emptyset, \emptyset)$$

For empty views PM_{la} and LM_{la}, none of the refinement rules (4.100)–(4.122) applies as none of their LHSs matches. Thus, the RHSs of the refinement rules imply empty PM_{la+1} and LM_{la+1}:

$$PM_{la+1} = (\emptyset, \emptyset, \emptyset, \emptyset, \emptyset)$$
$$LM_{la+1} = (\emptyset, \emptyset, \emptyset, \emptyset, \emptyset, \emptyset, \emptyset)$$

These are again valid views of PM and LM.

Induction Hypothesis 2

Let (4.130) hold for particular la, DM, MM, $PM_{highestLayer}$, $LM_{highestLayer}$, PM_{la}, and LM_{la}, i.e. valid views $PM_{highestLayer}$ and $LM_{highestLayer}$ were refined by (4.100)–(4.122) into valid views PM_{la} and LM_{la}.

Induction Step 2

We now extend PM_{la} and LM_{la} to valid views PM'_{la} and LM'_{la} in any possible way. We then refine PM'_{la} and LM'_{la} by (4.100)–(4.122) into PM'_{la+1} and LM'_{la+1}. Finally, we show that PM'_{la+1} and LM'_{la+1} are still valid views of PM and LM, i.e. that (4.130) still holds. This is illustrated in figure 4.11.

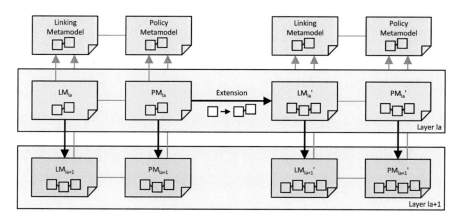

Figure 4.11: Induction step 2

1. Extension with a policy, so that PM'_{la} is still valid:

$$Policy'_{la} = Policy_{la} \cup (la, po', act') \tag{4.131}$$

In this case, (4.100) is triggered. This rule refines (la, po', act') and extends $Policy_{la+1}$ with a refined policy $(la + 1, po'', act')$ as follows:

$$\overset{(4.131)}{\Longrightarrow} \quad (la, po', act') \in Policy'_{la} \wedge la < lowestLayer$$

$$\overset{(4.100)}{\Longrightarrow} \quad \exists po''.(la + 1, po'', act') \in Policy'_{la+1} \wedge po' \xrightarrow[la \to la+1]{Po} po''$$

The refinement is complete as the refined tuple $(la + 1, po'', act')$ and the generated trace record do not cause any further refinement. In particular, (4.100) is not triggered again as the refined tuple cannot be contained in $Policy'_{la}$. (4.101), (4.102), and (4.104) are not triggered as po' is a new policy and therefore does not yet have any events, conditions, or actions. Thus:

$$Policy'_{la+1} = Policy_{la+1} \cup (la + 1, po'', act') \tag{4.132}$$

$$po' \xrightarrow[la \to la+1]{Po} po'' \tag{4.133}$$

This extension preserves the syntactical correctness. However, semantical correctness has to be checked, particularly (3.69). This constraint has been fulfilled without the tuple (induction hypothesis) and is still fulfilled with it, as shown in section B.1. Any other constraints are not affected and therefore still fulfilled.

Thus, PM'_{la+1} is a valid view of PM.

Furthermore, $LM'_{la+1} = LM_{la+1}$. Thus, LM'_{la+1} is a valid view of LM.

2. Extension with an event, so that PM'_{la} is still valid:

$$Event'_{la} = Event_{la} \cup (po', ev') \tag{4.134}$$

The refinement of the event ev' depends on whether ev' belongs to a policy that has already been refined, as described in (2a)–(2b).

a) ev' belongs to a policy and a trace record for the refinement of this policy exists:

$$po' \xrightarrow[la \to la+1]{Po} po'' \tag{4.135}$$

In this case, (4.101) is triggered. This rule refines (po', ev') and extends $Event_{la+1}$ with a refined event (po'', ev'') as follows:

$$\overset{(4.135),(4.134)}{\Longrightarrow} \quad po' \xrightarrow[la \to la+1]{Po} po'' \wedge (po', ev') \in Event'_{la}$$

$$\overset{(4.101)}{\Longrightarrow} \quad \exists ev''.(po'', ev'') \in Event'_{la+1} \wedge ev' \xrightarrow[la \to la+1]{Ev} ev''$$

The refinement is complete as the refined tuple (po'', ev'') and the generated trace record do not cause any further refinement. In particular, (4.101) is not triggered again as the refined tuple cannot be contained in $Event'_{la}$. Thus:

$$Event'_{la+1} = Event_{la+1} \cup (po'', ev'') \tag{4.136}$$

$$ev' \xrightarrow[la \to la+1]{Ev} ev'' \tag{4.137}$$

This extension preserves the syntactical correctness. However, semantical correctness has to be checked, particularly (3.76). This constraint has been fulfilled without the tuple (induction hypothesis) and is still fulfilled with it, as shown in section B.2. Any other constraints are not affected and therefore still fulfilled.

Thus, PM'_{la+1} is a valid view of PM.

b) If (2a) does not apply, no refinement rule is triggered. In this case, $PM'_{la+1} = PM_{la+1}$. Thus, PM'_{la+1} is a valid view of PM.

In both cases 2a and 2b, $LM'_{la+1} = LM_{la+1}$. Thus, LM'_{la+1} is a valid view of LM.

3. Extension with a condition, so that PM'_{la} is still valid:

$$Condition'_{la} = Condition_{la} \cup (po', cd', ty') \tag{4.138}$$

The refinement of the condition cd' depends on whether cd' belongs to a policy that has already been refined and whether cd' is nested as inner condition in an outer condition that has already been refined, as described in (3a)–(3b).

a) cd' belongs to a policy and a trace record for the refinement of this policy exists:

$$po' \xrightarrow[la \to la+1]{Po} po'' \tag{4.139}$$

In this case, (4.102) is triggered. This rule refines (po', cd', ty') and extends $Condition_{la+1}$ with a refined condition (po'', cd'', ty') as follows:

$$\begin{aligned} \overset{(4.139),(4.138)}{\Longrightarrow} \quad & po' \xrightarrow[la \to la+1]{Po} po'' \wedge (po', cd', ty') \in Condition'_{la} \\ \overset{(4.102)}{\Longrightarrow} \quad & \exists cd_2.(po'', cd'', ty') \in Condition'_{la+1} \wedge cd' \xrightarrow[la \to la+1]{Cd} cd'' \end{aligned}$$

The refinement is complete as the refined tuple (po'', cd'', ty') and the generated trace record do not cause any further refinement. In particular, (4.102) is not triggered again as the refined tuple cannot be contained in $Condition'_{la}$. If cd'' is contained as outer condition in a tuple in $ConditionNesting'_{la}$, (4.103) might be triggered again; in this case, however, the tuple must already have been contained in $ConditionNesting_{la}$ and therefore already have been refined in the induction hypothesis. The same applies to (4.106)–(4.115); in these cases, the respective tuples must already have been refined in the induction hypothesis. Thus:

$$Condition'_{la+1} = Condition_{la+1} \cup (po'', cd'', ty') \tag{4.140}$$

$$cd' \xrightarrow[la \to la+1]{Cd} cd'' \tag{4.141}$$

This extension preserves the syntactical correctness. However, semantical correctness has to be checked, particularly (3.85)–(3.86). These constraints have been fulfilled without the tuple (induction hypothesis) and are still fulfilled with it, as shown in section B.3. Any other constraints are not affected and therefore still fulfilled.

Thus, PM'_{la+1} is a valid view of PM.

b) If (3a) does not apply, no refinement rule is triggered. In this case, $PM'_{la+1} = PM_{la+1}$. Thus, PM'_{la+1} is a valid view of PM.

In both cases 3a and 3b, $LM'_{la+1} = LM_{la+1}$. Thus, LM'_{la+1} is a valid view of LM.

4. Extension with a condition nesting, so that PM'_{la} is still valid:

$$ConditionNesting'_{la} = ConditionNesting_{la} \cup (cd'_2, cd'_1) \tag{4.142}$$
$$Condition'_{la} = Condition_{la} \tag{4.143}$$

The refinement of the condition nesting (cd'_2, cd'_1) and the inner condition cd'_1 depend on whether the outer condition cd'_2 has already been refined, as described in (4a)–(4b).

a) A trace record for the refinement of cd'_2 exists:

$$cd'_2 \xrightarrow[la \to la+1]{Cd} cd''_2 \tag{4.144}$$

In this case, (4.103) is triggered. This rule refines (cd'_2, cd'_1) and cd_1 and extends $ConditionNesting_{la+1}$ with a refined nesting (cd''_2, cd''_1) and $Condition_{la+1}$ with a refined condition $(undef, cd'_1, ty')$ as follows:

$$\overset{4.142}{\Longrightarrow} (cd'_2, cd'_1) \in ConditionNesting'_{la}$$
$$\overset{I.H.}{\Longrightarrow} \exists ty'.(undef, cd'_1, ty') \in Condition'_{la} \tag{4.145}$$

$$\overset{\substack{(4.144),(4.142),\\(4.145)}}{\Longrightarrow} cd'_2 \xrightarrow[la \to la+1]{Cd} cd''_2 \wedge (cd'_2, cd'_1) \in ConditionNesting'_{la}$$
$$\wedge (undef, cd'_1, ty') \in Condition'_{la}$$
$$\overset{(4.103)}{\Longrightarrow} \exists cd''_1.(cd''_2, cd''_1) \in ConditionNesting'_{la+1}$$
$$\wedge (undef, cd''_1, ty') \in Condition'_{la+1} \wedge cd'_1 \xrightarrow[la \to la+1]{Cd} cd''_1$$

The refinement is complete as the refined tuples (cd_2'', cd_1'') and $(undef, cd_1'', ty')$ and the generated trace record do not cause any further refinement. In particular, (4.102) is not triggered as $(undef, cd_1'', ty')$ does not refer to any policy. If cd_1'' is contained as outer condition in a tuple in $ConditionNesting_{la'}'$ (4.103) might be triggered again; in this case, however, the tuple must already have been contained in $ConditionNesting_{la}$ and therefore have been refined in the induction hypothesis. The same applies to (4.106)–(4.115); in these cases, the respective tuples must already have been refined in the induction hypothesis. Thus:

$$ConditionNesting_{la+1}' = ConditionNesting_{la+1} \tag{4.146}$$
$$\cup\, (cd_2'', cd_1'')$$

$$Condition_{la+1}' = Condition_{la+1} \cup (undef, cd_1'', ty') \tag{4.147}$$

$$cd_1' \xrightarrow[\;la \to la+1\;]{Cd} cd_1'' \tag{4.148}$$

If cd_1' is also contained in another outer condition, it has already been refined into cd_1''. In this case, the refined tuple $(undef, cd_1'', ty')$ does already exist in $Condition_{la+1}$, and $Condition_{la+1}$ is not extended. Otherwise, $Condition_{la+1}$ is extended. This extension preserves the syntactical correctness. However, semantical correctness has to be checked, particularly (3.85)–(3.86) and (3.96)–(3.101). These constraints have been fulfilled without the tuples (induction hypothesis) and are still fulfilled with them, as shown in section B.4. Any other constraints are not affected and therefore still fulfilled.

Thus, PM_{la+1}' is a valid view of PM.

b) If (3a) does not apply, no refinement rule is triggered. In this case, $PM_{la+1}' = PM_{la+1}$. Thus, PM_{la+1}' is a valid view of PM.

In both cases 4a and 4b, $LM_{la+1}' = LM_{la+1}$. Thus, LM_{la+1}' is a valid view of LM.

5. Extension with an action, so that PM'_{la} is still valid:

$$Action'_{la} = Action_{la} \cup (po', ac', no') \qquad (4.149)$$

The refinement of the action ac' depends on whether ac' belongs to a policy that has already been refined, as described in (5a)–(5b).

    a) ac' belongs to a policy and a trace record for the refinement of this policy exists:

$$po' \xrightarrow[la \to la+1]{Po} po'' \qquad (4.150)$$

In this case, (4.104) is triggered. This rule refines (po', ac', no') and extends $Action_{la+1}$ with a refined action (po'', ac'', no'') as follows:

$$\overset{(4.150),(4.149)}{\Longrightarrow} \quad po' \xrightarrow[la \to la+1]{Po} po'' \wedge (po', ac', no') \in Action'_{la}$$

$$\overset{(4.104)}{\Longrightarrow} \quad \exists ac''.(po'', ac'', no') \in Action'_{la+1} \wedge ac' \xrightarrow[la \to la+1]{Ac} ac''$$

The refinement is complete as the refined tuple (po'', ac'', no') and the generated trace record do not cause any further refinement. In particular, (4.104) is not triggered again as the refined tuple cannot be contained in $Action'_{la}$. Thus:

$$Action'_{la+1} = Action_{la+1} \cup (po'', ac'', no') \qquad (4.151)$$

$$ac' \xrightarrow[la \to la+1]{Ac} ac'' \qquad (4.152)$$

This extension preserves the syntactical correctness. However, semantical correctness has to be checked, particularly (3.106)–(3.107). These constraints have been fulfilled without the tuple (induction hypothesis) and are still fulfilled with it, as shown in section B.5. Any other constraints are not affected and therefore still fulfilled.

Thus, PM'_{la+1} is a valid view of PM.

    b) If (5a) does not apply, no refinement rule is triggered. In this case, $PM'_{la+1} = PM_{la+1}$. Thus, PM'_{la+1} is a valid view of PM.

In both cases 5a and 5b, $LM'_{la+1} = LM_{la+1}$. Thus, LM'_{la+1} is a valid view of LM.

6. Extension with an event link, so that PM'_{la} is still valid:

$$EL'_{la} = EL_{la} \cup (ev', co') \tag{4.153}$$

The refinement of the event link (ev', co') depends on whether the enclosed event ev' has already been refined and a mapping for the enclosed concept co' was modeled, as described in (6a)–(6b).

a) A trace record for the refinement of ev' and a mapping for co' exist:

$$ev' \xrightarrow[la \to la+1]{Ev} ev'' \tag{4.154}$$

$$co' \xrightarrow[la \to la+1]{Co} (co''_1, ..., co''_n) \tag{4.155}$$

In this case, (4.105) is triggered. This rule refines (ev', co') and extends EL_{la+1} with a refined event link (ev'', co''_i) as follows:

$$\overset{(4.154),(4.155),}{\underset{(4.153)}{\Longrightarrow}} \quad ev' \xrightarrow[la \to la+1]{Ev} ev'' \wedge co' \xrightarrow[la \to la+1]{Co} (co''_1, ..., co''_n)$$
$$\wedge (ev', co') \in EL'_{la}$$

$$\overset{(4.105)}{\Longrightarrow} \quad \bigoplus_{i \in \{1,...,n\} \wedge co''_i \neq undef} (ev'', co''_i) \in EL'_{la+1}$$

The refinement is complete as the refined tuple (ev_2, co_{2_i}) does not cause any further refinement. If the refined tuple is also contained in EL'_{la}, (4.105) might be triggered again; in this case, however, the refined tuple must already have been contained in EL_{la} and therefore have been refined in the induction hypothesis. Thus:

$$EL'_{la+1} = EL_{la+1} \cup (ev'', co''_i) \wedge co''_i \neq undef \tag{4.156}$$

This extension preserves the syntactical correctness. However, semantical correctness has to be checked, particularly (3.134)–(3.136). These constraints have been fulfilled without the tuple (induction hypothesis) and are still fulfilled with it, as shown in section B.6. Any other constraints are not affected and therefore still fulfilled.

Thus, LM'_{la+1} is a valid view of LM.

b) If (6a) does not apply, no refinement rule is triggered. In this case, $LM'_{la+1} = LM_{la+1}$. Thus, LM'_{la+1} is a valid view of LM.

In both cases 6a and 6b, $PM'_{la+1} = PM_{la+1}$. Thus, PM'_{la+1} is a valid view of PM.

7. Extension with a binary expression link, so that PM'_{la} is still valid:

$$BEL'_{la} = BEL_{la} \cup (be', arg'_1, arg'_2) \tag{4.157}$$

The refinement of the binary expression link (be', arg'_1, arg'_2) depends on whether the enclosed binary expression be' has already been refined and, depending on the enclosed arguments arg'_1 and arg'_2, whether mappings for the arguments were modeled or the arguments have already been refined, as described in (7a)–(7j).

a) A trace record for the refinement of be' exists and arg'_1 and arg'_2 are literals:

$$be' \xrightarrow[la \to la+1]{Cd} be'' \tag{4.158}$$

$$arg'_1 \in Lit \wedge arg'_2 \in Lit \tag{4.159}$$

In this case, (4.106) is triggered. This rule refines (be', arg'_1, arg'_2) and extends BEL_{la+1} with a refined binary expression link (be'', arg'_1, arg'_2) as follows:

$\overset{(4.158),(4.157),}{\underset{\Longrightarrow}{(4.159)}}$ $be' \xrightarrow[la \to la+1]{Cd} be'' \wedge (be', arg'_1, arg'_2) \in BEL'_{la}$

$\qquad\qquad \wedge arg'_1 \in Lit \wedge arg'_2 \in Lit$

$\overset{(4.106)}{\Longrightarrow}$ $(be'', arg'_1, arg'_2) \in BEL'_{la+1}$

The refinement is complete as the refined tuple (be'', arg'_1, arg'_2) does not cause any further refinement. If the refined tuple is also contained in BEL'_{la}, (4.106) might be triggered again; in this case, however, the refined tuple must already have been contained in BEL_{la} and therefore have been refined in the induction hypothesis. Thus:

$$BEL'_{la+1} = BEL_{la+1} \cup (be'', arg'_1, arg'_2) \tag{4.160}$$

This extension preserves the syntactical correctness. However, semantical correctness has to be checked, particularly (3.143)–(3.147). These constraints have been fulfilled without the tuple (induction hypothesis) and are still fulfilled with it, as shown in section B.7. Any other constraints are not affected and therefore still fulfilled.

Thus, LM'_{la+1} is a valid view of LM.

b) A trace record for the refinement of be' exists, arg_1' and arg_2' are properties, and mappings for both exist:

$$be' \xrightarrow[la \to la+1]{Cd} be'' \tag{4.161}$$

$$arg_1' \in Id_{Pr} \wedge arg_2' \in Id_{Pr} \tag{4.162}$$

$$arg_1' \xrightarrow[la \to la+1]{Pr} (arg_{1_1}'', ..., arg_{1_m}'') \tag{4.163}$$

$$arg_2' \xrightarrow[la \to la+1]{Pr} (arg_{2_1}'', ..., arg_{2_n}'') \tag{4.164}$$

In this case, (4.107) is triggered. This rule refines (be', arg_1', arg_2') and extends BEL_{la+1} with a refined binary expression link $(be'', arg_{1_i}'', arg_{2_j}'')$ as follows:

$$\begin{array}{c} (4.161),(4.163),(4.164), \\ (4.157),(4.162) \\ \Longrightarrow \end{array} \qquad be' \xrightarrow[la \to la+1]{Cd} be'' \wedge arg_1' \xrightarrow[la \to la+1]{Pr} (arg_{1_1}'', ..., arg_{1_m}'')$$

$$\wedge arg_2' \xrightarrow[la \to la+1]{Pr} (arg_{2_1}'', ..., arg_{2_n}'')$$

$$\wedge (be', arg_1', arg_2') \in BEL_{la}' \wedge arg_1' \in Id_{Pr} \wedge arg_2' \in Id_{Pr}$$

$$\begin{array}{c} (4.107) \\ \Longrightarrow \end{array} \qquad \bigoplus_{\substack{i \in \{1,...,m\} \wedge arg_{1_i}'' \neq undef \\ \wedge j \in \{1,...,n\} \wedge arg_{2_j}'' \neq undef}} (be', arg_{1_i}'', arg_{2_j}'') \in BEL_{la+1}'$$

The refinement is complete as the refined tuple $(be'', arg_{1_i}'', arg_{2_j}'')$ does not cause any further refinement. If the refined tuple is also contained in BEL_{la}', (4.107) might be triggered again; in this case, however, the refined tuple must already have been contained in BEL_{la} and therefore have been refined in the induction hypothesis. Thus:

$$BEL_{la+1}' = BEL_{la} \cup (be'', arg_{1_i}'', arg_{2_j}'')$$
$$\wedge arg_{1_i}'' \neq undef \wedge arg_{2_j}'' \neq undef \tag{4.165}$$

This extension preserves the syntactical correctness. However, semantical correctness has to be checked, particularly (3.143)–(3.147). These constraints have been fulfilled without the tuple (induction hypothesis) and are still fulfilled with it, as shown in section B.7. Any other constraints are not affected and therefore still fulfilled.

Thus, LM_{la+1}' is a valid view of LM.

c) A trace record for the refinement of be' exists and arg'_1 and arg'_2 are operation invokings:

$$be' \xrightarrow[la \to la+1]{Cd} be'' \tag{4.166}$$

$$arg'_1 \in Id_{OI} \wedge arg'_2 \in Id_{OI} \tag{4.167}$$

In this case, (4.108) is triggered. This rule refines (be', arg'_1, arg'_2), arg'_1, and arg'_2 and extends BEL_{la+1} with a refined binary expression link (be'', arg''_1, arg''_2) as follows:

(4.166),(4.157),

(4.167)

$$\Longrightarrow \qquad be' \xrightarrow[la \to la \mid 1]{Cd} be'' \wedge (be', arg'_1, arg'_2) \in BEL'_{la}$$

$$\wedge arg'_1 \in Id_{OI} \wedge arg'_2 \in Id_{OI}$$

(4.108)

$$\Longrightarrow \quad \exists arg''_1, arg''_2.(be'', arg''_1, arg''_2) \in BEL'_{la+1}$$

$$\wedge arg'_1 \xrightarrow[la \to la+1]{OI} arg''_1 \wedge arg'_2 \xrightarrow[la \to la+1]{OI} arg''_2$$

The refined tuple $(be'', arg''_{1_i}, arg''_2)$ does not cause any further refinement. If the refined tuple is also contained in BEL'_{la}, (4.108) might be triggered again; in this case, however, the refined tuple must already have been contained in BEL_{la} and therefore have been refined in the induction hypothesis. Thus:

$$BEL'_{la+1} = BEL_{la+1} \cup (be'', arg''_1, arg''_2) \tag{4.168}$$

$$arg'_1 \xrightarrow[la \to la+1]{OI} arg''_1 \tag{4.169}$$

$$arg'_2 \xrightarrow[la \to la+1]{OI} arg''_2 \tag{4.170}$$

This extension preserves the syntactical correctness. However, semantical correctness has to be checked, particularly (3.143)–(3.147). These constraints have been fulfilled without the tuples (induction hypothesis) and are still fulfilled with them, as shown in section B.7. Any other constraints are not affected and therefore still fulfilled.

The generated trace records might cause further refinements as they trigger (4.117) to (4.122). If arg'_1 and arg'_2 have already been refined due to an occurrence elsewhere, $OIL1_{la+1}$, $OIL2_{la+1}$, and $OIL3_{la+1}$ are not extended; otherwise, they might be extended with additional tuples. The respective cases 11 and 12 show that (3.120)–(3.123) are still fulfilled after the refinement.

Thus, LM'_{la+1} is a valid view of LM.

d) A trace record for the refinement of be' exists, arg'_1 is a property, arg'_2 is a literal, and a mapping for arg'_1 exists:

$$be' \xrightarrow[la \to la+1]{Cd} be'' \tag{4.171}$$

$$arg'_1 \in Id_{Pr} \wedge arg'_2 \in Lit \tag{4.172}$$

$$arg'_1 \xrightarrow[la \to la+1]{Pr} (arg''_{1_1}, ..., arg''_{1_n}) \tag{4.173}$$

In this case, (4.109) is triggered. This rule refines (be', arg'_1, arg'_2) and extends BEL_{la+1} with a refined binary expression link $(be'', arg''_{1_i}, arg''_2)$ as follows:

$(4.171),(4.173),$
$(4.157),(4.172)$
$\Longrightarrow$
$$be' \xrightarrow[la \to la+1]{Cd} be'' \wedge arg'_1 \xrightarrow[la \to la+1]{Pr} (arg''_{1_1}, ..., arg''_{1_n})$$
$$\wedge (be', arg'_1, arg'_2) \in BEL'_{la} \wedge arg'_1 \in Id_{Pr} \wedge arg'_2 \in Lit$$

(4.109)
$\Longrightarrow$
$$\exists arg''_2. \bigoplus_{i \in \{1,...,n\} \wedge arg''_{1_i} \neq undef} (be'', arg''_{1_i}, arg''_2) \in BEL'_{la+1}$$
$$\wedge arg''_2 \in mappedLiterals^{Pr}_{la \to la+1}(arg'_1, arg'_2, arg''_{1_i})$$

The refinement is complete as the refined tuple $(be'', arg''_{1_i}, arg''_2)$ does not cause any further refinement. If the refined tuple is also contained in BEL'_{la}, (4.109) might be triggered again; in this case, however, the refined tuple must already have been contained in BEL_{la} and therefore have been refined in the induction hypothesis. Thus:

$$BEL'_{la+1} = BEL_{la+1} \cup (be'', arg''_{1_i}, arg''_2) \wedge arg''_{1_i} \neq undef \tag{4.174}$$

$$arg''_2 \in mappedLiterals^{Pr}_{la \to la+1}(arg'_1, arg'_2, arg''_{1_i}) \tag{4.175}$$

This extension preserves the syntactical correctness. However, semantical correctness has to be checked, particularly (3.143)–(3.147). These constraints have been fulfilled without the tuple (induction hypothesis) and are still fulfilled with it, as shown in section B.7. Any other constraints are not affected and therefore still fulfilled.

Thus, LM'_{la+1} is a valid view of LM.

e) A trace record for the refinement of be' exists, arg_1' is a literal, arg_2' is a property, and a mapping for arg_2' exists:

$$be' \xrightarrow[la \to la+1]{Cd} be'' \tag{4.176}$$

$$arg_1' \in Lit \land arg_2' \in Id_{Pr} \tag{4.177}$$

$$arg_2' \xrightarrow[la \to la+1]{Pr} (arg_{2_1}'', ..., arg_{2_n}'') \tag{4.178}$$

In this case, (4.110) is triggered. This rule refines (be', arg_1', arg_2') and extends BEL_{la+1} with a refined binary expression link $(be'', arg_1'', arg_{2_i}'')$ as follows:

(4.176),(4.178),
(4.157),(4.177)
$\Longrightarrow$

$$be' \xrightarrow[la \to la+1]{Cd} be'' \land arg_2' \xrightarrow[la \to la+1]{Pr} (arg_{2_1}'', ..., arg_{2_n}'')$$
$$\land (be', arg_1', arg_2') \in BEL_{la}' \land arg_1' \in Lit \land arg_2' \in Id_{Pr}$$

(4.110)
$\Longrightarrow$

$$\exists arg_1''. \bigoplus_{i \in \{1,...,n\} \land arg_{2_i}'' \neq undef} (be'', arg_1'', arg_{2_i}'') \in BEL_{la+1}'$$
$$\land arg_1'' \in mappedLiterals_{la \to la+1}^{Pr}(arg_1', arg_2', arg_{2_i}'')$$

The refinement is complete as the refined tuple $(be'', arg_1'', arg_{2_i}'')$ does not cause any further refinement. If the refined tuple is also contained in BEL_{la}', (4.110) might be triggered again; in this case, however, the refined tuple must already have been contained in BEL_{la} and therefore have been refined in the induction hypothesis. Thus:

$$BEL_{la+1}' = BEL_{la+1} \cup (be'', arg_1'', arg_{2_i}'') \land arg_{2_i}'' \neq undef \tag{4.179}$$

$$arg_1'' \in mappedLiterals_{la \to la+1}^{Pr}(arg_1', arg_2', arg_{2_i}'') \tag{4.180}$$

This extension preserves the syntactical correctness. However, semantical correctness has to be checked, particularly (3.143)–(3.147). These constraints have been fulfilled without the tuple (induction hypothesis) and are still fulfilled with it, as shown in section B.7. Any other constraints are not affected and therefore still fulfilled.

Thus, LM_{la+1}' is a valid view of LM.

f) A trace record for the refinement of be' exists, arg_1' is an operation invoking, and arg_2' is a literal:

$$be' \xrightarrow[la \to la+1]{Cd} be'' \tag{4.181}$$

$$arg_1' \in Id_{OI} \wedge arg_2' \in Lit \tag{4.182}$$

In this case, (4.111) is triggered. This rule refines (be', arg_1', arg_2') and arg_1' and extends BEL_{la+1} with a refined binary expression link (be'', arg_1'', arg_2'') as follows:

$$(4.181),(4.157),$$
$$(4.182)$$
$$\Longrightarrow \quad be' \xrightarrow[la \to la+1]{Cd} be'' \wedge (be', arg_1', arg_2') \in BEL_{la}'$$

$$\wedge arg_1' \in Id_{OI} \wedge arg_2' \in Lit$$

$$\overset{(4.111)}{\Longrightarrow} \quad \exists arg_1'', arg_2''.(be'', arg_1'', arg_2'') \in BEL_{la+1}'$$

$$\wedge arg_2'' \in mappedLiterals_{la \to la+1}^{Op}(operationOfOpInv(arg_1'),$$

$$arg_2', operationOfOpInv(arg_1'')) \wedge arg_1' \xrightarrow[la \to la+1]{OI} arg_1''$$

The refined tuple (be'', arg_1'', arg_2'') does not cause any further refinement. If the refined tuple is also contained in BEL_{la}', (4.111) might be triggered again; in this case, however, the refined tuple must already have been contained in BEL_{la} and therefore have been refined in the induction hypothesis. Thus:

$$BEL_{la+1}' = BEL_{la+1} \cup (be'', arg_1'', arg_2'') \tag{4.183}$$

$$arg_2'' \in mappedLiterals_{la \to la+1}^{Op}(operationOfOpInv$$
$$(arg_1'), arg_2', operationOfOpInv(arg_1'')) \tag{4.184}$$

$$arg_1' \xrightarrow[la \to la+1]{OI} arg_1'' \tag{4.185}$$

This extension preserves the syntactical correctness. However, semantical correctness has to be checked, particularly (3.143)–(3.147). These constraints have been fulfilled without the tuples (induction hypothesis) and are still fulfilled with them, as shown in section B.7. Any other constraints are not affected and therefore still fulfilled.

The generated trace record might cause further refinements as it triggers (4.117) to (4.122). If arg_1' has already been refined due to an occurrence elsewhere, $OIL1_{la+1}$, $OIL2_{la+1}$, and $OIL3_{la+1}$ are not extended; otherwise, they might be extended with additional tuples. The respective cases 11 and 12 show that (3.120)–(3.123) are still fulfilled after the refinement.

Thus, LM_{la+1}' is a valid view of LM.

g) A trace record for the refinement of be' exists, arg_1' is a literal, and arg_2' is an operation invoking:

$$be' \xrightarrow[la \to la+1]{Cd} be'' \tag{4.186}$$

$$arg_1' \in Lit \wedge arg_2' \in Id_{OI} \tag{4.187}$$

In this case, (4.112) is triggered. This rule refines (be', arg_1', arg_2') and arg_2' and extends BEL_{la+1} with a refined binary expression link (be'', arg_1'', arg_2'') as follows:

$(4.186),(4.157),$

(4.187)

$\implies$

$$be' \xrightarrow[la \to la+1]{Cd} be'' \wedge (be', arg_1', arg_2') \in BEL_{la}'$$

$$\wedge \, arg_1' \in Lit \wedge arg_2' \in Id_{OI}$$

(4.112)

$\implies$

$$\exists arg_1'', arg_2''. (be'', arg_1'', arg_2'') \in BEL_{la+1}'$$

$$\wedge \, arg_1'' \in mappedLiterals_{la \to la+1}^{Op}(operationOfOpInv(arg_2'),$$

$$arg_1', operationOfOpInv(arg_2'')) \wedge arg_2' \xrightarrow[la \to la+1]{OI} arg_2''$$

The refined tuple (be'', arg_1'', arg_2'') does not cause any further refinement. If the refined tuple is also contained in BEL_{la}', (4.112) might be triggered again; in this case, however, the refined tuple must already have been contained in BEL_{la} and therefore have been refined in the induction hypothesis. Thus:

$$BEL_{la+1}' = BEL_{la+1} \cup (be'', arg_1'', arg_2'') \tag{4.188}$$

$$arg_1'' \in mappedLiterals_{la \to la+1}^{Op}(operationOfOpInv \\ (arg_2'), arg_1', operationOfOpInv(arg_2'')) \tag{4.189}$$

$$arg_2' \xrightarrow[la \to la+1]{OI} arg_2'' \tag{4.190}$$

This extension preserves the syntactical correctness. However, semantical correctness has to be checked, particularly (3.143)–(3.147). These constraints have been fulfilled without the tuples (induction hypothesis) and are still fulfilled with them, as shown in section B.7. Any other constraints are not affected and therefore still fulfilled.

The generated trace record might cause further refinements as it triggers (4.117) to (4.122). If arg_2' has already been refined due to an occurrence elsewhere, $OIL1_{la+1}$, $OIL2_{la+1}$, and $OIL3_{la+1}$ are not extended; otherwise, they might be extended with additional tuples. The respective cases 11 and 12 show that (3.120)–(3.123) are still fulfilled after the refinement.

Thus, LM_{la+1}' is a valid view of LM.

h) A trace record for the refinement of be' exists, arg_1' is an operation invoking, arg_2' is a property, and a mapping for arg_2' exists:

$$be' \xrightarrow[la \to la+1]{Cd} be'' \tag{4.191}$$

$$arg_1' \in Id_{OI} \wedge arg_2' \in Id_{Pr} \tag{4.192}$$

$$arg_2' \xrightarrow[la \to la+1]{Pr} (arg_{2_1}'', ..., arg_{2_n}'') \tag{4.193}$$

In this case, (4.113) is triggered. This rule refines (be', arg_1', arg_2') and arg_1' and extends BEL_{la+1} with a refined binary expression link (be'', arg_1'', arg_2'') as follows:

$(4.191),(4.193),$
$(4.157)(4.192)$
$\implies$
$$be' \xrightarrow[la \to la+1]{Cd} be'' \wedge arg_2' \xrightarrow[la \to la+1]{Pr} (arg_{2_1}'', ..., arg_{2_n}'')$$
$$\wedge (be', arg_1', arg_2') \in BEL_{la}' \wedge arg_1' \in Id_{OI} \wedge arg_2' \in Id_{Pr}$$

(4.113)
$\implies$
$$\exists arg_1''. \bigoplus_{i \in \{1,...,n\} \wedge arg_{2_i}'' \neq undef} (be'', arg_1'', arg_{2_i}'') \in BEL_{la+1}'$$
$$\wedge arg_1' \xrightarrow[la \to la+1]{OI} arg_1''$$

The refined tuple $(be'', arg_1'', arg_{2_i}'')$ does not cause any further refinement. If the refined tuple is also contained in BEL_{la}', (4.113) might be triggered again; in this case, however, the refined tuple must already have been contained in BEL_{la} and therefore have been refined in the induction hypothesis. Thus:

$$BEL_{la+1}' = BEL_{la+1} \cup (be'', arg_1'', arg_{2_i}'') \wedge arg_{2_i}'' \neq undef \tag{4.194}$$

$$arg_1' \xrightarrow[la \to la+1]{OI} arg_1'' \tag{4.195}$$

This extension preserves the syntactical correctness. However, semantical correctness has to be checked, particularly (3.143)–(3.147). These constraints have been fulfilled without the tuples (induction hypothesis) and are still fulfilled with them, as shown in section B.7. Any other constraints are not affected and therefore still fulfilled.

The generated trace record might cause further refinements as it triggers (4.117) to (4.122). If arg_1' has already been refined due to an occurrence elsewhere, $OIL1_{la+1}$, $OIL2_{la+1}$, and $OIL3_{la+1}$ are not extended; otherwise, they might be extended with additional tuples. The respective cases 11 and 12 show that (3.120)–(3.123) are still fulfilled after the refinement.

Thus, LM_{la+1}' is a valid view of LM.

i) A trace record for the refinement of be' exists, arg'_1 is a property, arg'_2 is an operation invoking, and a mapping for arg'_1 exists:

$$be' \xrightarrow[la \to la+1]{Cd} be'' \tag{4.196}$$

$$arg'_1 \in Id_{Pr} \wedge arg'_2 \in Id_{OI} \tag{4.197}$$

$$arg'_1 \xrightarrow[la \to la+1]{Pr} (arg''_{1_1}, ..., arg''_{1_n}) \tag{4.198}$$

In this case, (4.114) is triggered. This rule refines (be', arg'_1, arg'_2) and arg'_2 and extends BEL_{la+1} with a refined binary expression link (be'', arg''_1, arg''_2) as follows:

$$\begin{array}{l} (4.196),(4.198), \\ (4.157),(4.197) \\ \implies \end{array} \quad be' \xrightarrow[la \to la+1]{Cd} be'' \wedge arg'_1 \xrightarrow[la \to la+1]{Pr} (arg''_{1_1}, ..., arg''_{1_n})$$

$$\wedge \, (be', arg'_1, arg'_2) \in BEL'_{la} \wedge arg'_1 \in Id_{Pr} \wedge arg'_2 \in Id_{OI}$$

$$\begin{array}{l} (4.114) \\ \implies \end{array} \exists arg''_2. \bigoplus_{i \in \{1,...,n\} \wedge arg''_{2_i} \neq undef} (be'', arg''_{1_i}, arg''_2) \in BEL'_{la+1}$$

$$\wedge \, arg'_2 \xrightarrow[lu \to lu+1]{OI} arg''_2$$

The refined tuple $(be'', arg''_{1_i}, arg''_2)$ does not cause any further refinement. If the refined tuple is also contained in BEL'_{la}, (4.114) might be triggered again; in this case, however, the refined tuple must already have been contained in BEL_{la} and therefore have been refined in the induction hypothesis. Thus:

$$BEL'_{la+1} = BEL_{la+1} \cup (be'', arg''_{1_i}, arg''_2) \wedge arg''_{1_i} \neq undef \tag{4.199}$$

$$arg'_2 \xrightarrow[la \to la+1]{OI} arg''_2 \tag{4.200}$$

This extension preserves the syntactical correctness. However, semantical correctness has to be checked, particularly (3.143)–(3.147). These constraints have been fulfilled without the tuples (induction hypothesis) and are still fulfilled with them, as shown in section B.7. Any other constraints are not affected and therefore still fulfilled.

The generated trace record might cause further refinements as it triggers (4.117) to (4.122). If arg'_2 has already been refined due to an occurrence elsewhere, $OIL1_{la+1}$, $OIL2_{la+1}$, and $OIL3_{la+1}$ are not extended; otherwise, they might be extended with additional tuples. The respective cases 11 and 12 show that (3.120)–(3.123) are still fulfilled after the refinement.

Thus, LM'_{la+1} is a valid view of LM.

j) If (7a) to (7i) do not apply, no refinement rule is triggered. In this case, $LM'_{la+1} = LM_{la+1}$. Thus, LM'_{la+1} is a valid view of LM.

In all cases 7a to 7j, $PM'_{la+1} = PM_{la+1}$. Thus, PM'_{la+1} is a valid view of PM.

8. Extension with an operational expression link, so that PM'_{la} is still valid:

$$OEL'_{la} = OEL_{la} \cup (oe', oi') \tag{4.201}$$

The refinement of the operational expression link (oe', oi') depends on whether the enclosed operational expression oe' and and operation invoking oi' have already been refined, as described in (8a)–(8b).

a) A trace record for the refinement of oe' exists:

$$oe' \xrightarrow[la \to la+1]{Cd} oe'' \tag{4.202}$$

In this case, (4.115) is triggered. This rule refines (oe', oi') and oi' and extends OEL_{la+1} with a refined operational expression link (oe'', oi'') as follows:

$$\overset{(4.202),(4.201)}{\Longrightarrow} \quad oe' \xrightarrow[la \to la+1]{Cd} oe''$$
$$\wedge (oe', oi') \in OEL'_{la}$$
$$\overset{(4.115)}{\Longrightarrow} \quad \exists oi''. (oe'', oi'') \in OEL'_{la+1}$$
$$\wedge oi' \xrightarrow[la \to la+1]{OI} oi''$$

The refined tuple (oe'', oi'') does not cause any further refinement. If the refined tuple is also contained in OEL'_{la}, (4.115) might be triggered again; in this case, however, the refined tuple must already have been contained in OEL_{la} and therefore have been refined in the induction hypothesis. Thus:

$$OEL'_{la+1} = OEL_{la+1} \cup (oe'', oi'') \tag{4.203}$$

$$oi' \xrightarrow[la \to la+1]{OI} oi'' \tag{4.204}$$

This extension preserves the syntactical correctness. However, semantical correctness has to be checked, particularly (3.156)–(3.159). These constraints have been fulfilled without the tuples (induction hypothesis) and are still fulfilled with them, as shown in section B.8. Any other constraints are not affected and therefore still fulfilled.

The generated trace record might cause further refinements as it triggers (4.117) to (4.122). If oi' has already been refined due to an occurrence elsewhere, $OIL1_{la+1}$, $OIL2_{la+1}$, and $OIL3_{la+1}$ are not extended; otherwise, they might be extended with additional tuples. The respective cases 11 and 12 show that (3.120)–(3.123) are still fulfilled after the refinement.

Thus, LM'_{la+1} is a valid view of LM.

b) If (8a) does not apply, no refinement rule is triggered. In this case, $LM'_{la+1} = LM_{la+1}$. Thus, LM'_{la+1} is a valid view of LM.

In both cases 8a and 8b, $PM'_{la+1} = PM_{la+1}$. Thus, PM'_{la+1} is a valid view of PM.

9. Extension with an action link, so that PM'_{la} is still valid:

$$AL'_{la} = AL_{la} \cup (ac', oi') \tag{4.205}$$

The refinement of the action link (ac', oi') depends on whether the enclosed action ac' and and operation invoking oi' have already been refined, as described in (9a)–(9b).

a) A trace record for the refinement of ac' exists:

$$ac' \xrightarrow[la \to la+1]{Ac} ac'' \tag{4.206}$$

In this case, (4.116) is triggered. This rule refines (ac', oi') and oi' and extends AL_{la+1} with a refined action link (ac'', oi'') as follows:

$$\overset{(4.206),(4.205)}{\Longrightarrow} \quad ac' \xrightarrow[la \to la+1]{Ac} ac''$$
$$\land (ac', oi') \in AL'_{la}$$

$$\overset{(4.116)}{\Longrightarrow} \quad \exists oi''.(ac'', oi'') \in AL'_{la+1}$$
$$\land oi' \xrightarrow[la \to la+1]{OI} oi''$$

The refined tuple (ac'', oi'') does not cause any further refinement. If the refined tuple is also contained in AL'_{la}, (4.116) might be triggered again; in this case, however, the refined tuple must already have been contained in AL_{la} and therefore have been refined in the induction hypothesis. Thus:

$$AL'_{la+1} = AL_{la+1} \cup (ac'', oi'') \tag{4.207}$$

$$oi' \xrightarrow[la \to la+1]{OI} oi'' \tag{4.208}$$

This extension preserves the syntactical correctness. However, semantical correctness has to be checked, particularly (3.163)–(3.166). These constraints have been fulfilled without the tuples (induction hypothesis) and are still fulfilled with them, as shown in section B.9. Any other constraints are not affected and therefore still fulfilled.

The generated trace record might cause further refinements as it triggers (4.117) to (4.122). If oi' has already been refined due to an occurrence elsewhere, $OIL1_{la+1}$, $OIL2_{la+1}$, and $OIL3_{la+1}$ are not extended; otherwise, they might be extended with additional tuples. The respective cases 11 and 12 show that (3.120)–(3.123) are still fulfilled after the refinement.

Thus, LM'_{la+1} is a valid view of LM.

b) If (8a) does not apply, no refinement rule is triggered. In this case, $LM'_{la+1} = LM_{la+1}$. Thus, LM'_{la+1} is a valid view of LM.

In both cases 9a and 9b, $PM'_{la+1} = PM_{la+1}$. Thus, PM'_{la+1} is a valid view of PM.

10. Extension with an operation invoking, so that PM'_{la} is still valid:

$$OIL1'_{la} = OIL1_{la} \cup (oi') \tag{4.209}$$
$$OIL2'_{la} = OIL2_{la} \tag{4.210}$$
$$OIL3'_{la} = OIL3_{la} \tag{4.211}$$

If a trace record for the refinement of oi' existed, (4.117) would be triggered. However, no trace record exists as oi' is a new operation invoking. Thus, no refinement rule is triggered. $LM'_{la+1} = LM_{la+1}$. Thus, LM'_{la+1} is a valid view of LM.

PM_{la+1} is not extended either and thus still a valid view of PM.

11. Extension with an operation invoking link, so that PM'_{la} is still valid:

$$OIL1'_{la} = OIL1_{la} \tag{4.212}$$
$$OIL2'_{la} = OIL2_{la} \cup (oi', op') \tag{4.213}$$
$$OIL3'_{la} = OIL3_{la} \tag{4.214}$$

The refinement of the operation invoking link (oi', op') depends on whether the enclosed operation invoking oi' has already been refined and a mapping for the enclosed operation op' was modeled, as described in (11a)–(11b).

a) A trace record for the refinement of oi' and a mapping for op' exist, and a set of parameters appears with the refined operation:

$$oi' \xrightarrow[la \to la+1]{OI} oi'' \tag{4.215}$$

$$op' \xrightarrow[la \to la+1]{Op} (op''_1, ..., op''_n) \tag{4.216}$$

$$\xrightarrow[la \to la+1]{Pa} (pa''_1, ..., pa''_n) \wedge \forall i \in \{1, ..., n\}.pa''_i \tag{4.217}$$
$$\in parametersOfOperation(operationOfOpInv(oi''))$$

In this case, (4.118) is triggered. This rule refines (oi', op') and extends $OIL2_{la+1}$ with a refined operation invoking link $(oi'', op2_i)$ as follows:

$$\begin{array}{c} (4.215),(4.216),\\ (4.213)\\ \Longrightarrow \end{array} \quad oi' \xrightarrow[la \to la+1]{OI} oi'' \wedge op' \xrightarrow[la \to la+1]{Op} (op''_1, ..., op''_n)$$
$$\wedge (oi', op') \in OIL2'_{la}$$

$$\begin{array}{c} (4.118)\\ \Longrightarrow \end{array} \quad \bigoplus_{i \in \{1,...,n\} \wedge op''_i \neq undef} (oi'', op''_i) \in OIL2'_{la+1}$$

If parameters appear with the refined operation, they are linked to the operation oi'' with one of their values:

$$\begin{array}{c} (4.215),(4.217)\\ \Longrightarrow \end{array} \quad oi' \xrightarrow[la \to la+1]{OI} oi'' \wedge \xrightarrow[la \to la+1]{Pa} (pa''_1, ..., pa''_n) \wedge \forall i \in \{1, ..., n\}.$$
$$pa''_i \in parametersOfOperation(operationOfOpInv(oi''))$$

$$\begin{array}{c} (4.122)\\ \Longrightarrow \end{array} \quad \exists lit''. \bigoplus_{i \in \{1,...,n\} \wedge pa''_i \neq undef \wedge (oi'', op'') \in OIL2_{la+1}} (oi'', pa''_i, lit'') \in OIL3'_{la+1}$$
$$\wedge op'' = operationOfParameter(pa''_i)$$
$$\wedge lit'' \in appearedLiterals^{Pa}_{la \to la+1}(pa''_i)$$

The refinement is complete as the refined tuples (oi'', op_i'') and (oi'', pa_i'', lit'') do not cause any further refinement. If the refined tuple is also contained in $OIL2'_{la}$, (4.118) might be triggered again; in this case, however, the refined tuple must already have been contained in $OIL2_{la}$ and therefore have been refined in the induction hypothesis. Thus:

$$OIL1'_{la+1} = OIL1_{la+1} \tag{4.218}$$
$$OIL2'_{la+1} = OIL2_{la+1} \cup (oi'', op_i'') \wedge op_i'' \neq undef \tag{4.219}$$
$$OIL3'_{la+1} = OIL3_{la+1} \cup (oi'', pa_i'', lit'') \wedge pa_i'' \neq undef$$
$$\wedge (oi'', operationOfParameter(pa_i'')) \in OIL2_{la+1} \tag{4.220}$$
$$lit'' \in appearedLiterals^{Pa}_{la \rightarrow la+1}(pa_i'') \tag{4.221}$$

This extension preserves the syntactical correctness. However, semantical correctness has to be checked, particularly (3.120)–(3.123). These constraints have been fulfilled without the tuple (induction hypothesis) and are still fulfilled with it, as shown in section B.10. Any other constraints are not affected and therefore still fulfilled.

Thus, LM'_{la+1} is a valid view of LM.

b) If (11a) does not apply, no refinement rule is triggered. In this case, $LM'_{la+1} = LM_{la+1}$. Thus, LM'_{la+1} is a valid view of LM.

In both cases 11a and 11b, $PM'_{la+1} = PM_{la+1}$. Thus, PM'_{la+1} is a valid view of PM.

12. Extension with an operation invoking link, so that PM'_{la} is still valid:

$$OIL1'_{la} = OIL1_{la} \tag{4.222}$$
$$OIL2'_{la} = OIL2_{la} \tag{4.223}$$
$$OIL3'_{la} = OIL3_{la} \cup (oi', pa', arg') \tag{4.224}$$

The refinement of the operation invoking link (oi', pa', arg') depends on whether the enclosed operation invoking oi' has already been refined, a mapping for the enclosed parameter pa' was modeled, and, depending on the enclosed argument arg', whether a mapping for the argument was modeled or the argument has already been refined, as described in (12a)–(12d).

a) A trace record for the refinement of oi' and a mapping for pa' exist and arg' is a literal:

$$oi' \xrightarrow[la \to la+1]{OI} oi'' \tag{4.225}$$

$$pa' \xrightarrow[la \to la+1]{Pa} (pa_1'', ..., pa_n'') \tag{4.226}$$

$$arg' \in Lit \tag{4.227}$$

In this case, (4.119) is triggered. This rule refines (oi', pa', arg_1') and extends $OIL3_{la+1}$ with a refined operation invoking link (oi'', pa_i'', arg_2'') as follows:

$$\begin{array}{l} \text{(4.225),(4.226),} \\ \text{(4.224),(4.227)} \\ \Longrightarrow \end{array} \quad oi' \xrightarrow[la \to la+1]{OI} oi'' \wedge pa' \xrightarrow[la \to la+1]{Pa} (pa_1'', ..., pa_n'')$$

$$\wedge (oi', pa', arg') \in OIL3_{la}' \wedge arg_1' \in Lit$$

$$\xRightarrow{\text{(4.119)}} \exists arg''. \qquad \bigoplus_{\substack{i \in \{1,...,n\} \wedge pa_i'' \neq undef \wedge (oi'',op'') \in OIL2_{la+1} \\ \wedge op'' = operationOfParameter(pa_i'')}} (oi'', pa_i'', arg'') \in OIL3_{la+1}'$$

$$\wedge arg'' \in mappedLiterals_{la \to la+1}^{Pa}(pa', arg', pa_i'')$$

The refinement is complete as the refined tuple (oi'', pa_i'', arg'') does not cause any further refinement. If the refined tuple is also contained in $OIL3_{la}'$, (4.119) might be triggered again; in this case, however, the refined tuple must already have been contained in $OIL3_{la}$ and therefore have been refined in the induction hypothesis. Thus:

$$OIL1_{la+1}' = OIL1_{la+1} \tag{4.228}$$

$$OIL2_{la+1}' = OIL2_{la+1} \tag{4.229}$$

$$OIL3_{la+1}' = OIL3_{la+1} \cup (oi'', pa_i'', arg'') \wedge pa_i'' \neq undef \tag{4.230}$$
$$\wedge (oi'', operationOfParameter(pa_i'')) \in OIL2_{la+1}$$

$$arg'' \in mappedLiterals_{la \to la+1}^{Pa}(pa', arg', pa_i'') \tag{4.231}$$

This extension preserves the syntactical correctness. However, semantical correctness has to be checked, particularly (3.120)–(3.123). These constraints have been fulfilled without the tuple (induction hypothesis) and are still fulfilled with it, as shown in section B.10. Any other constraints are not affected and therefore still fulfilled.

Thus, LM_{la+1}' is a valid view of LM.

b) A trace record for the refinement of oi' and a mapping for pa' exist, arg' is a property, and a mapping for arg' exists:

$$oi' \xrightarrow[la \rightarrow la+1]{OI} oi'' \tag{4.232}$$

$$pa' \xrightarrow[la \rightarrow la+1]{Pa} (pa''_1, ..., pa''_m) \tag{4.233}$$

$$arg' \in Id_{Pr} \tag{4.234}$$

$$arg' \xrightarrow[la \rightarrow la+1]{Pr} (arg''_1, ..., arg''_n) \tag{4.235}$$

In this case, (4.120) is triggered. This rule refines (oi', pa', arg') and extends $OIL3_{la+1}$ with a refined operation invoking link (oi'', pa''_i, arg''_j) as follows:

$$\begin{array}{l} \text{(4.232),(4.233),(4.235),} \\ \text{(4.224),(4.234)} \\ \Longrightarrow \end{array} \qquad oi' \xrightarrow[la \rightarrow la+1]{OI} oi'' \wedge pa' \xrightarrow[la \rightarrow la+1]{Pa} (pa''_1, ..., pa''_m)$$

$$\wedge\, arg' \xrightarrow[la \rightarrow la+1]{Pr} (arg''_1, ..., arg''_n)$$

$$\wedge\, (oi', pa', arg') \in OIL3'_{la} \wedge arg' \in Id_{Pr}$$

$$\begin{array}{l} \text{(4.120)} \\ \Longrightarrow \end{array} \qquad \bigoplus_{\substack{i \in \{1,...,m\} \wedge pa''_i \neq undef \wedge (oi'',op'') \in OIL2_{la+1} \\ \wedge op'' = operationOfParameter(pa''_i) \\ \wedge j \in \{1,...,n\} \wedge arg''_j \neq undef}} \qquad (oi'', pa''_i, arg''_j) \in OIL3'_{la+1}$$

The refinement is complete as the refined tuple (oi'', pa''_i, arg''_j) does not cause any further refinement. If the refined tuple is also contained in $OIL3'_{la}$, (4.120) might be triggered again; in this case, however, the refined tuple must already have been contained in $OIL3_{la}$ and therefore have been refined in the induction hypothesis. Thus:

$$OIL1'_{la+1} = OIL1_{la+1} \tag{4.236}$$

$$OIL2'_{la+1} = OIL2_{la+1} \tag{4.237}$$

$$OIL3'_{la+1} = OIL3_{la+1} \cup (oi'', pa''_i, arg''_j) \wedge pa''_i \neq undef$$
$$\wedge\, (oi'', operationOfParameter(pa''_i)) \in OIL2_{la+1} \tag{4.238}$$
$$\wedge\, arg''_j \neq undef$$

This extension preserves the syntactical correctness. However, semantical correctness has to be checked, particularly (3.120)–(3.123). These constraints have been fulfilled without the tuple (induction hypothesis) and are still fulfilled with it, as shown in section B.10. Any other constraints are not affected and therefore still fulfilled.

Thus, LM'_{la+1} is a valid view of LM.

c) A trace record for the refinement of oi' and a mapping for pa' exist and arg' is an operation invoking:

$$oi' \xrightarrow[la \to la+1]{OI} oi'' \tag{4.239}$$

$$pa' \xrightarrow[la \to la+1]{Pa} (pa''_1, ..., pa''_n) \tag{4.240}$$

$$arg' \in Id_{OI} \tag{4.241}$$

In this case, (4.121) is triggered. This rule refines (oi', pa', arg') and arg' and extends $OIL3_{la+1}$ with a refined operation invoking link (oi'', pa''_i, arg'') as follows:

$$\begin{aligned}
&\overset{(4.239),(4.240),}{\underset{\Longrightarrow}{(4.224),(4.241)}} \quad oi' \xrightarrow[la \to la+1]{OI} oi'' \wedge pa' \xrightarrow[la \to la+1]{Pa} (pa''_1, ..., pa''_n) \\
&\qquad\qquad\qquad \wedge (oi', pa', arg') \in OIL3'_{la} \wedge arg' \in Id_{OI} \\[2mm]
&\overset{(4.121)}{\Longrightarrow} \quad \exists arg''. \qquad\qquad \bigoplus_{\substack{i \in \{1,...,n\} \wedge pa''_i \neq undef \wedge (oi'',op'') \in OIL2_{la+1} \\ \wedge op''=operationOfParameter(pa''_i)}} \qquad (oi'', pa''_i, arg'') \in OIL3'_{la+1} \\[2mm]
&\qquad\qquad\qquad \wedge arg' \xrightarrow[la \to la+1]{OI} arg''
\end{aligned}$$

The refined tuple (oi'', pa''_i, arg'') does not cause any further refinement. If the refined tuple is also contained in $OIL3'_{la}$, (4.121) might be triggered again; in this case, however, the refined tuple must already have been contained in $OIL3_{la}$ and therefore have been refined in the induction hypothesis.

Thus:

$$OIL1'_{la+1} = OIL1_{la+1} \tag{4.242}$$

$$OIL2'_{la+1} = OIL2_{la+1} \tag{4.243}$$

$$OIL3'_{la+1} = OIL3_{la+1} \cup (oi'', pa''_i, arg'') \wedge pa''_i \neq undef$$
$$\wedge (oi'', operationOfParameter(pa''_i)) \in OIL2_{la+1} \tag{4.244}$$

$$arg' \xrightarrow[la \to la+1]{OI} arg'' \tag{4.245}$$

This extension preserves the syntactical correctness. However, semantical correctness has to be checked, particularly (3.120)–(3.123). These constraints have been fulfilled without the tuples (induction hypothesis) and are still fulfilled with them, as shown in section B.10. Any other constraints are not affected and therefore still fulfilled.

The generated trace record might cause further refinements as it triggers (4.117) to (4.122). If oi'' has already been refined due to an occurrence elsewhere, $OIL1_{la+1}$, $OIL2_{la+1}$, and $OIL3_{la+1}$ are not extended; otherwise, they might be extended with additional tuples. The respective cases 11 and 12 show that (3.120)–(3.123) are still fulfilled after the refinement.

Thus, LM'_{la+1} is a valid view of LM.

d) If (12a) to (12c) do not apply, no refinement rule is triggered. In this case, $LM'_{la+1} = LM_{la+1}$. Thus, LM'_{la+1} is a valid view of LM.

In all cases 12a to 12d, $PM'_{la+1} = PM_{la+1}$. Thus, PM'_{la+1} is a valid view of PM.

In cases 1 to 12, we extended PM_{la} and LM_{la} to valid views PM'_{la} and LM'_{la} in any possible way. Then, we refined PM'_{la} and LM'_{la} with (4.100)–(4.122) into PM'_{la+1} and LM'_{la+1}. Finally, we showed that PM'_{la+1} and LM'_{la+1} were still valid views of PM and LM, i.e. that (4.130) still holds.

$\square$

4.2.4 Example

Previously, we provided an example scenario (cf. section 3.1) and specified the policies of this scenario with a policy and a linking model at the first abstraction layer (cf. sections 3.4.3 and 3.5.3). We also provided a mapping model for the respective domain (cf. section 4.1.4). These three models represent the input for the refinement rules. Now, we apply the refinement rules and thus generate the refined policy and linking models at the second abstraction layer. We demonstrate the different steps of the refinement in the following.

Policy Model

At first, the refinement rule for policies (definition 48) applies:

$$(1, lowQuality, true) \in Policy_1 \wedge 1 < 2$$

$$\overset{(4.100)}{\Rightarrow} (2, lowQuality2, true) \in Policy_2$$

$$\wedge\, lowQuality \xrightarrow[1\to2]{Po} lowQuality2$$

$$(1, highQuality, true) \in Policy_1 \wedge 1 < 2$$

$$\overset{(4.100)}{\Rightarrow} (2, highQuality2, true) \in Policy_2$$

$$\wedge\, highQuality \xrightarrow[1\to2]{Po} highQuality2$$

The refined policies are named *lowQuality2* and *highQuality2* and inserted into the policy model, and trace records for these refinements are added. These trace records trigger the refinement rule for events (definition 49), conditions (definition 50), and actions (definition 51):

$$lowQuality \xrightarrow[1\to2]{Po} lowQuality2$$

$$\wedge\, (lowQuality, lqEvent) \in Event_1$$

$$\overset{(4.101)}{\Rightarrow} (lowQuality2, lqEvent2) \in Event_2$$

$$\wedge\, lqEvent \xrightarrow[1\to2]{Ev} lqEvent2$$

$$highQuality \xrightarrow[1\to2]{Po} highQuality2$$

$$\wedge\, (highQuality, hqEvent) \in Event_1$$

$$\overset{(4.101)}{\Rightarrow} (highQuality2, hqEvent2) \in Event_2$$

$$\wedge\, hqEvent \xrightarrow[1\to2]{Ev} hqEvent2$$

$$lowQuality \xrightarrow[1 \to 2]{Po} lowQuality2$$

$$\land (lowQuality, lqCondition) \in Condition_1$$

$$\overset{(4.102)}{\Rightarrow} (lowQuality2, lqCondition2) \in Condition_2$$

$$\land lqCondition \xrightarrow[1 \to 2]{Cd} lqCondition2$$

$$highQuality \xrightarrow[1 \to 2]{Po} highQuality2$$

$$\land (highQuality, hqCondition) \in Condition_1$$

$$\overset{(4.102)}{\Rightarrow} (highQuality2, hqCondition2) \in Condition_2$$

$$\land hqCondition \xrightarrow[1 \to 2]{Cd} hqCondition2$$

$$lowQuality \xrightarrow[1 \to 2]{Po} lowQuality2$$

$$\land (lowQuality, lqAction) \in Action_1$$

$$\overset{(4.104)}{\Rightarrow} (lowQuality2, lqAction2) \in Action_2$$

$$\land lqAction \xrightarrow[1 \to 2]{Ac} lqAction2$$

$$highQuality \xrightarrow[1 \to 2]{Po} highQuality2$$

$$\land (highQuality, hqAction) \in Action_1$$

$$\overset{(4.104)}{\Rightarrow} (highQuality2, hqAction2) \in Action_2$$

$$\land hqAction \xrightarrow[1 \to 2]{Ac} hqAction2$$

The refined events are named *lqEvent2* and *hqEvent2*, the refined conditions are named *lqCondition2* and *hqCondition2*, and the refined actions are named *lqAction2* and *hqAction2*. All of them are inserted into the policy model, and trace records for these refinements are added. These trace records trigger the refinement rule for nested conditions (4.103), but no nested conditions were modeled.

Figure 4.12 shows an overview of the refined policy model with a UML-based concrete syntax. Tables 4.6 to 4.10 provide the formal representation of the refined policy model.

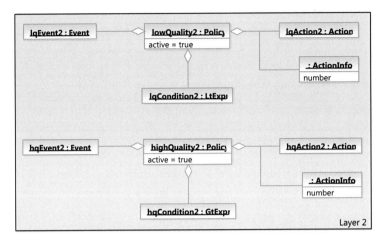

Figure 4.12: Example policy model (layer 2)

Table 4.6: Relation *Policy₂*

Layer	Name	Active
2	lowQuality2	true
2	highQuality2	true

Table 4.7: Relation *Event₂*

Policy	Name
lowQuality2	lqEvent2
highQuality2	hqEvent2

Table 4.8: Relation *Condition₂*

Policy	Name	Type
lowQuality2	lqCondition2	lt
highQuality2	hqCondition2	gt

Table 4.9: Relation *ConditionNesting₂*

OuterCondition	InnerCondition

Table 4.10: Relation *Action₂*

Policy	Name	No
lowQuality2	lqAction2	1
highQuality2	hqAction2	1

Linking Model

Now, the trace records for the refined events trigger the refinement rule for event links (definition 52):

$$lqEvent \xrightarrow[1\to2]{Ev} lqEvent2 \wedge signalQuality \xrightarrow[1\to2]{Co} intensityChange$$
$$\wedge\, (lqEvent, signalQuality) \in EL_1$$
$$\overset{(4.105)}{\Rightarrow}\, (lqEvent2, intensityChange) \in EL_2$$
$$hqEvent \xrightarrow[1\to2]{Ev} hqEvent2 \wedge signalQuality \xrightarrow[1\to2]{Co} intensityChange$$
$$\wedge\, (hqEvent, signalQuality) \in EL_1$$
$$\overset{(4.105)}{\Rightarrow}\, (hqEvent2, intensityChange) \in EL_2$$

The refined event links are inserted into the linking model. This refinement rule adds no trace record, so that no further refinement rules are triggered.

The trace records for the refined conditions trigger the refinement rules for binary expression links (definition 53):

$$lqCondition \xrightarrow[1\to2]{Cd} lqCondition2 \wedge sqNewValue \xrightarrow[1\to2]{Pr} icNewValue$$
$$\wedge\, (lqCondition, sqNewValue, 50) \in BEL_1$$
$$\overset{(4.109)}{\Rightarrow}\, (lqCondition2, icNewValue, 50) \in BEL_2$$
$$hqCondition \xrightarrow[1\to2]{Cd} hqCondition2 \wedge sqNewValue \xrightarrow[1\to2]{Pr} icNewValue$$
$$\wedge\, (hqCondition, sqNewValue, 80) \in BEL_1$$
$$\overset{(4.109)}{\Rightarrow}\, (hqCondition2, icNewValue, 80) \in BEL_2$$

The refined binary expression links are inserted into the linking model. This refinement rule adds no trace record, so that no further refinement rules are triggered. The trace records for the refined conditions also trigger the refinement rules for operational expression links (definition 54), but no operational expressions are modeled.

The trace records for the refined actions trigger the refinement rules for action links (definition 55):

$$lqAction \xrightarrow[1 \to 2]{Ac} lqAction2$$

$\wedge (lqAction, lqActionOpInv) \in AL_1$

$\overset{(4.116)}{\Rightarrow} (lqAction2, lqActionOpInv2) \in AL_2$

$\wedge lqActionOpInv \xrightarrow[1 \to 2]{OI} lqActionOpInv2$

$$hqAction \xrightarrow[1 \to 2]{Ac} hqAction2$$

$\wedge (hqAction, hqActionOpInv) \in AL_1$

$\overset{(4.116)}{\Rightarrow} (hqAction2, hqActionOpInv2) \in AL_2$

$\wedge hqActionOpInv \xrightarrow[1 \to 2]{OI} hqActionOpInv2$

The refined action links are inserted into the linking model. The refined operation invokings are named $lqActionOpInv2$ and $hqActionOpInv2$ and trace records for these refinements are added. These trace records trigger the refinement rule for operation invoking links (definition 56):

$$lqActionOpInv \xrightarrow[1 \to 2]{OI} lqActionOpInv2$$

$\wedge (lqActionOpInv) \in OIL1_1$

$\overset{(4.117)}{\Rightarrow} (lqActionOpInv2) \in OIL1_2$

$$hqActionOpInv \xrightarrow[1 \to 2]{OI} hqActionOpInv2$$

$\wedge (hqActionOpInv) \in OIL1_1$

$\overset{(4.117)}{\Rightarrow} (hqActionOpInv2) \in OIL1_2$

$$lqActionOpInv \xrightarrow[1 \to 2]{OI} lqActionOpInv2 \wedge increasePower \xrightarrow[1 \to 2]{Op} changeTXP$$

$\wedge (lqActionOpInv, increasePower) \in OIL2_1$

$\overset{(4.118)}{\Rightarrow} (lqActionOpInv2, changeTXP) \in OIL2_2$

$$hqActionOpInv \xrightarrow[1 \to 2]{OI} hqActionOpInv2 \wedge decreasePower \xrightarrow[1 \to 2]{Op} changeTXP$$

$\wedge (hqActionOpInv, decreasePower) \in OIL2_1$

$\overset{(4.118)}{\Rightarrow} (hqActionOpInv2, changeTXP) \in OIL2_2$

$$lqActionOpInv \xrightarrow[1 \to 2]{OI} lqActionOpInv2 \wedge \xrightarrow[1 \to 2]{Pa} ctChangeValue$$

$$\wedge \, ctChangeValue \in parametersOfOperation(changeTXP)$$

$$\overset{(4.122)}{\Rightarrow} (lqActionOpInv2, ctChangeValue, undef) \in OIL3_2$$

$$hqActionOpInv \xrightarrow[1 \to 2]{OI} hqActionOpInv2 \wedge \xrightarrow[1 \to 2]{Pa} ctChangeValue$$

$$\wedge \, ctChangeValue \in parametersOfOperation(changeTXP)$$

$$\overset{(4.122)}{\Rightarrow} (hqActionOpInv2, ctChangeValue, undef) \in OIL3_2$$

The refined operation invoking links are inserted into the linking model. This refinement rule adds no trace record, so that no further refinement rules are triggered.

Figure 4.13 shows an overview of the refined linking model with a UML-based concrete syntax. Tables 4.11 to 4.17 provide the formal representation of the refined linking model.

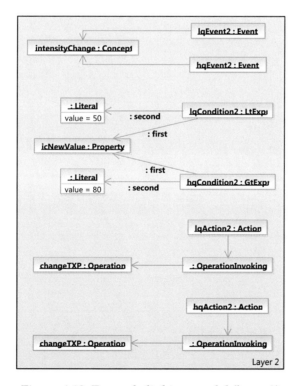

Figure 4.13: Example linking model (layer 2)

Table 4.11: Relation EL_2

Event	Concept
lqEvent2	intensityChange
hqEvent2	intensityChange

Table 4.12: Relation AL_2

Action	OpInv
lqAction2	lqActionOpInv2
hqAction2	hqActionOpInv2

Table 4.13: Relation BEL_2

BinExpr	Arg1	Arg2
lqCondition2	icNewValue	50
hqCondition2	icNewValue	80

Table 4.14: Relation OEL_2

OpExpr	OpInv

Table 4.15: Relation $OIL1_2$

OpInv
lqActionOpInv2
hqActionOpInv2

Table 4.16: Relation $OIL2_2$

OpInv	Operation
lqActionOpInv2	changeTXP
hqActionOpInv2	changeTXP

Table 4.17: Relation $OIL3_2$

OpInv	Parameter	Argument
lqActionOpInv2	ctChangeValue	undef
hqActionOpInv2	ctChangeValue	undef

4.3 Related Work

We now summarize typical approaches to the refinement of policies that are comparable to our approach. We first describe each approach and finally outline their commonalities and differences in a comparison table. Moreover, we compare them to the features of our approach.

SLA Policies

The authors of [118] present a refinement approach that focuses on policies within the autonomic networking domain. Policies represent configuration settings and are used for automated network and resource configuration. The approach is tailored to the implementation of an example scenario, in which an Internet service provider offers different connection services to customers: Voice over IP (VoIP), Video on Demand (VoD), and Web. Depending on the respective Service Level Agreement (SLA), each customer is granted a particular total bandwidth according to the customer's service package (gold, silver, or bronze). Policies are used to manage bandwidth partitioning in order to satisfy the SLA statements. For this purpose, customers assign grades (1, 2, or 3) to their different connection services. The total bandwidth of a customer is then partitioned amongst the customer's services depending on the service package and the service grades. As a result, more or less bandwidth is allocated to the customer's VoIP, VoD, and Web connections.

A simple policy language is provided that offers a set of language objects. These objects can be composed to domain-specific policies for the example scenario. Policies are represented at different levels of a policy continuum. Each level provides a dialect of the policy language, which offers a sub-set of all language objects. A high-level policy e.g. expresses: "A customer has selected the gold service package with service grade 1 for VoIP connections". At a lower level, this policy is mapped to the network architecture, expressing: "Provide Differentiated Services (DiffServ) configuration for 6 Mbps with 50% of bandwidth for service grade 1".

The separation of domain-specific terminology from the actual policies allows to fully automate the refinement process. High-level business policies are refined into low-level device policies, as illustrated in figure 4.14. A policy wizard provides a Graphical User Interface (GUI) to specify policies at the highest level, which is called the business level. The customer specifies policies to select a particular service package, to select the desired connection services, and to attach a service grade to each selected service. For this purpose, dropdown menus contain the language objects to be selected by the customer. These dropdown menus cover all terminology needed at the

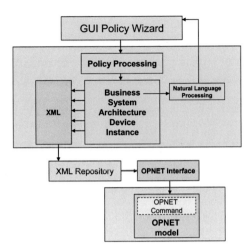

Figure 4.14: Refinement of bandwidth policies [118]

business level. Once a policy is specified, an internal Extensible Markup Language (XML)-based representation of the policy is generated for further processing. Afterwards, XML-based representations are generated for the lower levels of the policy continuum until the lowest level is reached, which is called the instance level. This level represents the policies as specific configuration commands on a per-device basis.

The prototype implementation uses a language parser to transform information from the policy wizard into the business dialect of the policy language. For the generation of the lower-level representations, XSL Transformations (XSLTs) [172] are used that replace the objects of the higher level with the respective objects at the subsequent lower level. An XML pipeline processes the transformations level per level.

Workflow Policies

In [125], a refinement approach is presented that derives low-level workflow policies from high-level ones using refinement patterns. A workflow represents a task sequence that is executed in a certain ordering. Workflow policies put constraints on a workflow by specifying properties that must hold in each task of the workflow when being executed. They can e.g. be used for managing security aspects of the workflow. However, no specific example or application are provided. Refinement transforms abstract policies referring to a workflow as a whole into a specific representation referring to the single tasks of the workflow by using different vocabulary and adding information. The used refinement patterns encapsulate domain knowledge, so that it can be applied to the workflow policies in an automated way.

The approach is restricted to deterministic workflows. In a deterministic workflow, a task can have different possible successors, but the selection of the actual successor during workflow execution must be unambiguous. A formalism is used that models each workflow as a deterministic state machine. In this formalism, each single task of the workflow is represented as a state. The sequence of state transitions during workflow execution represents the execution trace of the workflow. The state machine is then transformed into a Kripke model in order to formalize the workflow. This Kripke model allows to characterize all possible execution traces within the workflow by means of a temporal logic formula. A workflow policy then specifies a property that must be fulfilled in any state of the workflow during its execution trace. A high-level workflow policy makes a statement about the workflow in an abstract way. A low-level representation of a high-level workflow policy policy makes the same statement, but expresses it with enforceable constraints about specific properties.

The refinement of workflow policies is performed in an iterative process and starts with the initial set of high-level workflow policies. In each step of the iteration, the refinement patterns are applied to the current set of workflow policies. For this purpose, a refinement pattern consists of the three parts context, problem, and solution. The context part describes to which tasks of the workflow the pattern can be applied. The problem part describes the abstract property that should be refined. The solution suggests some less abstract properties or even specific constraints to replace the abstract property. Temporal logic is used to formally define the refinement patterns, and they operate on the temporal logic representation of the workflows. The applicable patterns are automatically selected by matching their context part to the execution flows and by matching their problem part to the properties used within the policies. As a result, the properties within the respective policies are replaced with the suggested properties. Thus, a set of less abstract policies is generated after each iteration and used as starting point for the next iteration. The refinement process is repeated until no more refinement pattern can be applied. This finally results in a set of low-level workflow constraints that satisfy the initial high-level workflow policies.

An implementation of the refinement approach is available for the generation of policies in the eXtensible Access Control Markup Language (XACML) [173]. The Protégé ontology editor [174] is used to specify workflow models and refinement patterns in the Web Ontology Language (OWL) [175]. The application of refinement patterns is implemented in Java, and the refinement tree is shown in a Java GUI. The XACML policy generation process is also implemented in Java.

Web Service Policies

In [126], an approach is presented that refines management policies for composite web services into enforceable policies for the constituent web services. In this context, policies are used for managing the behavior of adaptive web services. The behavior of a web service can be modified by changing the determining policies, and this makes the web service adaptive in a sense. An adaptive composite web service is one that is composed of other adaptive web services. The behavior of an adaptive composite service usually results from the behavior of the constituent adaptive services. This processes is now reversed, starting with high-level management polices that specify the desired behavior of an adaptive composite web service. The respective low-level polices are derived from these high-level management polices in order to enforce the desired behavior at the constituent web services. Instead of manually deriving low-level polices, the approach enables their automated generation. An example is presented roughly, in which a composite service generates report letters by composing three constituent services that collect data and perform the report generation.

The approach relies on a rich description of the web services and their behavior. The Web Services Description Language (WSDL) [176] is used to describe functional aspects of the service such as interfaces and parameters. Semantic Markup for Web Services (OWL-S) [177] is used to provide a semantic description by relating inputs and outputs of the service to definitions of an ontology. A Finite State Machine (FSM) provides a formal description of the adaptive service behavior by relating events, states, and state transitions to each other. This does not describe the entire internal logic of a service, but covers the parts of service behavior that can be adapted. Apart from service descriptions, the refinement is also based on the description of the composition. The approach further uses OWL-S to describe in detail how constituent web services are composed to a composite web service. This description also aggregates the adaptive behaviors of the constituent services to formally represent the behavior of the composite service in one large FSM. Compositions can be nested, allowing to further compose composite web services to larger composite services. Only linear compositions can be specified, and each composition results in a sequential execution of the services. Policies represent choices in the behavior of a web service and are expressed as Event Condition Action (ECA) rules. The provided example uses a high-level policy to specify the document type if the report is generated for a particular user. The event represents the request for report generation, the condition only considers the particular user, and the action is used to set the document type to a certain type.

The refinement is performed from top to bottom in a hierarchical way, starting with the policy specified for the uppermost service within the composition. This high-level policy specifies the adaptive behavior of the uppermost service. The description of the composition is used to refine the condition and action parts of the policy separately from each other. The condition refers to parameters in the composite service. According to the service's data flow, the condition is refined into conditions that refer to the respective parameters of the constituent services. The action controls the behavior of the composite service. According to the service's control flow, the action is refined into actions that control the respective behavior of the constituent services. As a result, low-level policies are generated for each constituent service that satisfy the original high-level policy. Thus, the refinement allows to modify the behavior of a composite web service without the need to redesign the service or to change the composition. More details about the policy refinement process are not provided by the authors.

A set of tools are available that support the specification of web services, web service compositions, and policies. These tools enable the creation of functionally correct policies without having in-depth expertise of web services or the refinement process. The refinement is implemented for Java web services. High-level and low-level policies are expressed in OWL and finally transformed into Jess rules [99] to be executed by a proprietary policy engine.

Monitoring Policies

The authors of [127] present a refinement approach that automatically derives low-level policies from high-level requirements based on measured data and classification techniques. High-level policies are regarded as requirements on the overall system behavior that have to be satisfied at a low level by the behavior of the system components. The system is considered a set of related entities, whereas each entity is characterized by a set of attributes. Low-level policies are in fact integrity rules that put constraints on the values of these low-level attributes. However, policies are not used to control the system components by setting attribute values, but are only used for monitoring purposes in order to determine whether the respective high-level requirements are fulfilled or not. If a policy is violated, it is the developer's task to solve the issue. An example is provided where low-level metrics are derived from high-level requirements on an auctioning platform. However, the approach claims to be independent of a particular domain.

Policies are represented at two abstraction levels. In the example, a high-level policy is represented by an SLA that specifies requirements on the system performance and availability: "With a maximum number of 400 clients connected to the auctioning platform, the throughput of the platform must be at least 17 auction requests per second, and each request must be answered in at most 24 seconds". At the low level, a number of technical system metrics such as Central Processing Unit (CPU) utilization are available. The objective is to derive which value ranges are allowed for these metrics, so that the high-level SLA is still satisfied. These ranges represent the respective low-level policies that monitor the system components. The CPU is one example component. A violation of a policy indicates an SLA violation.

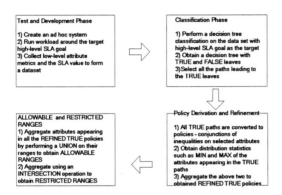

Figure 4.15: Refinement of performance policies [127]

The policy refinement process breaks down the high-level SLAs into smaller low-level policies and is partitioned into three phases. First, simulation runs are executed with the system in different scenarios by placing different workloads on the system. In the example, simulations regard different numbers of auction clients and different numbers of auction requests. During the simulation, the system is monitored and data is collected on both the low-level metrics and the fulfillment of the high-level SLA. Metrics that do not significantly vary for different workloads are eliminated. The second phase analyzes the collected data and uses classification algorithms to determine for which combinations of values of the metrics the SLA is fulfilled. The appropriate value ranges are extracted and specified as constraints on the metrics. In the third phase, different value ranges for the same metrics are combined. For each metric, an appropriate low-level policy is derived that specifies which value range satisfies the SLA, such as: "CPU usage should remain between 1.08% and 12.434%".

However, the approach is not always accurate and may result in incorrectly classified attributes as classification is based on measured data. The results depend on the collected data and on the used classification algorithm. Different tree classification algorithms are analyzed and result in different attributes to be considered significant, in different value ranges, and in different error rates.

Goal Policies

In [128, 129], a refinement approach is presented that allows to infer low-level actions from high-level goals. The focus is on computing the sequence of low-level operations that are available in an underlying system and that satisfy a high-level goal when being executed. Actually, goals are refined into a set of operations that achieve them. However, the initial goals are regarded as high-level policies, and the inferred operations are regarded as action clauses of the respective low-level policies that achieve the initial goals. Thus, policy refinement is an application and extension of the goal refinement approach. The approach is illustrated by an example from the network management domain. In this example, an enterprise network must fulfill an SLA that assures a particular Quality of Service (QoS) for an application. The objective is inferring operations of a router to change the network configuration accordingly.

Goals exist at different levels of abstraction in the approach. An abstract goal uses high-level terms that are not related directly to system components. The initial SLA goal in the example is formulated at a high level: "On demand, the network should be configured to provide Gold QoS to the web service application". In contrast, a specific goal is one that is assigned to a specific system component whose capabilities enable the system to satisfy the goal, such as: "The parameters of the DiffServ router should be set correctly". The approach provides a UML profile for modeling the goals. Each goal contains an attribute for its internal representation, which has to be specified manually by the user. Any goal is internally represented as a temporal logic rule. The initial SLA goal can e.g. be specified as: "If a packet is sent to the web service application, then this packet is eventually assigned Gold QoS".

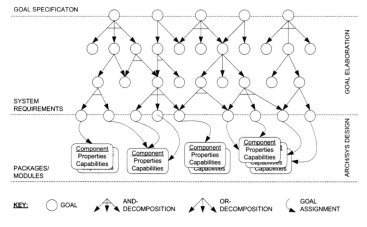

Figure 4.16: Refinement of goal policies [128]

The refinement approach is divided into two phases. The first phase is decomposing the initial abstract goals into specific subgoals that logically entail the initial ones. Finding subgoals is a manual task, and two types of decomposition are available for this: *and* means that any subgoal must be achieved, whereas *or* means that only one of the subgoals must be achieved. The UML profile also provides a high-level notation for goal decomposition. Refinement patterns are available for the different decomposition types. The high-level notation for the decomposition is then automatically transformed into the respective temporal logic representation. One example is the decomposition of one goal A into two subgoals B and C, whereas B implies that eventually C is true, and C implies that eventually A is true. The specification of goals and goal refinements result in a formally defined goal refinement hierarchy. In the example, the abstract SLA goal would be decomposed into a set of subgoals that describe the configuration changes to be applied to a router in the network.

The second phase in the refinement approach takes the formally decomposed goals and automatically identifies the objects and operations in the system architecture that will achieve them. For this purpose, additional information is modeled as follows. First, a domain hierarchy represents the relationships between abstract entities and specific system objects and allows to map the abstract entities used in high-level goals to these system objects. Second, the underlying system is specified in terms of system objects and their behavior. For each system object, a UML class diagram specifies its parameters and operations, and a UML state chart describes the pre- and postconditions of its operations. The domain hierarchy and the UML diagrams are transformed automatically into a formal representation. A reasoning mechanism finally compares the system specification to the decomposed goals. It will find the required sequence of operations to achieve each subgoal, if a solution exists. In the example, the inferred actions perform the relevant configuration changes at a router to provide Gold QoS to the web service application. The resulting sequence of operations can be used as the action clause of a policy to achieve this respective subgoal. Thus, low-level policies are inferred that satisfy the subgoals. If all subgoals are covered by appropriate policies, the initial goals are finally satisfied.

The entire formalism is formalized in Event Calculus (EC), and any specified information is transformed into EC. In the first phase, the logical representation of the goals is transformed into EC as a prerequisite for the second phase. Additionally, EC allows to validate the goal decomposition and to check it for inconsistencies. In the second phase, EC provides means for abductive reasoning, which is used to infer the sequence of operations to achieve a goal. Tool support is not yet available, but the integration of a UML editor is planned to specify system behavior and goal information in UML and to translate these specifications into EC for further processing.

Comparison

The presented approaches to the refinement of policies differ in some aspects from our approach. We outline the differences in table 4.18. A + in the table indicates that a feature is available in an approach, a o indicates a limited feature, and a - indicates that a feature is not available at all.

Table 4.18: Comparison of typical policy refinement approaches

	Bandwidth policies [118]	Workflow policies [125]	Web service policies [126]	Performance policies [127]	Goal policies [128, 129]	Model-Driven Policy Refinement
ECA policies	-	-	+	-	-	+
Language-independent	+	+	+	+	+	+
Customizable domain	-	-	-	+	+	+
Abstraction levels	o	+	+	o	o	+
Automation	+	+	+	+	o	+
Interaction possible	-	-	-	-	+	+
Formal semantics	-	+	-	-	-	+

The presented approaches support different policy types. Besides our approach, only the one for web service policies addresses ECA policies. This approach represents ECA policies with XML, whereas we represent them with a relational algebra. The approach for bandwidth policies also represents policies with XML, but addresses configuration policies in the form of attribute-value pairs. The approach for workflow policies addresses constraint policies and represents them as temporal logic formulas. The approach for performance policies also addresses constraint policies, but represents them as attribute-value pairs. The approach for goal policies uses goals as input policies, represents them as temporal logic formulas, and generates actions as output policies. All approaches are independent of any policy language or engine and enable the refinement of policies in a language-independent way.

The domain of the policies cannot be customized in the approaches for bandwidth, workflow, and web service policies as these are tailored to one particular domain. In contrast, a customized domain is possible in the approach for performance policies by adapting the classification. In the approach for goal policies, the domain can be customized by means of the system description and and the mapping of abstract system entities to specific system objects. This is similar to the specification of the domain model and the mappings used in our approach.

Policies are considered at different abstraction levels in all approaches. The minimal number of levels that allows for a refinement is two, and this is the number of levels in the approaches for performance and goal policies. The approach for bandwidth policies also uses a fixed number of abstraction levels to represent policies, but enables five different levels. As in our approach, the approaches for workflow and web service policies enable a flexible number of abstraction layers.

All approaches except the one for goal policies are fully automated and generate refined policies. In the latter one, the refinement of goals into subgoals is a manual process, and operations are inferred automatically from the refined subgoals afterwards. Thus, this approach is only semi-automated. In exchange, the approach for goal policies is the only one that allows for interaction with the developer during the refinement process, apart from our approach. The formal semantics of the refinement process is only provided in the approach for workflow policies and in our approach. A prerequisite for the definition of formal semantics is a formal policy representation. Although the approach for workflow policies does represent policies in a formal way, it does not provide a detailled description of the refinement process.

4.4 Conclusion and Outlook

The way policy development is performed today is contrary to the paradigms of modern software development as application policies are decoupled from their technical representation and even more from their implementation. The manual refinement and implementation imposes additional effort. As a solution to this issue, we presented a domain-specific approach to the automated refinement of ECA policies in this chapter that makes use of model-driven techniques. Our approach makes high-level application policies executable by refinement into a machine-executable representation at a technical level. The refinement process is automated with mapping patterns and refinement rules that replace higher-layer objects with lower-layer ones and thus generate the refined models. The refined policies at the lowest abstraction layer represent the starting point for an automated transformation of the respective models into executable code for purposes of implementation.

Our approach transfers benefits of Model-Driven Engineering (MDE) to Policy-Based Management (PBM) in several ways. The usage of models enables the specification of policies at a high level of abstraction initially and avoids their direct implementation at a technical level. Models do not only serve documentation purposes, but they are essential artifacts of the policy development process. Policies are refined from the highest to the lowest layer in an automated way, starting with non-executable application policies and resulting in a technical representation of these policies. We achieve this with model transformations, which are an essential part of any effective model-driven solution. The automated refinement helps to consolidate development effort and allows to control the actual system behavior by changing the high-level models, and this is even possible at runtime.

The degree of refinement automation highly depends on an effective specification of the domain and the mappings, which are the essential input for the refinement rules. Our approach also respects manual interaction, which can often not be avoided completely. However, manual steps as reduced as far as possible, and whenever a refinement step cannot be performed automatically, the possible solutions are proposed. The available mapping patterns defined by the mapping metamodel determine the expressiveness of the refinement. We use mappings that replace an object or a combination of objects with another object or combination of objects. Furthermore, we only allow for mappings between subsequent layers in order to avoid ambiguities. Otherwise, an object might be derived from different higher layers in different ways. If complex dependencies should be modeled that cannot be specified with the available mappings patterns, the mapping metamodel must be extended in order to enable a more expressive set of mappings and refinement rules.

We define the formal semantics of the refinement process with a relational algebra and use this algebra to prove that valid models are refined into valid models again, i.e. that the refinement rules preserve the static semantics of the models. Valid models are a prerequisite for the generation of executable code. The possibility to generate code eliminates the dependency on a particular policy language as the same models can be used to generate code for various languages. The feasibility of code generation for PonderTalk has already been shown [42]. Our approach is not only generic with respect to the language, but also with respect to the domain and the number of abstraction layers, and it is nevertheless automated. All working solutions known to us are either tailored to a particular problem or domain, or they do not realize an automated refinement of policies, or they do not define formal execution semantics.

An effective concrete syntax for the mappings is being developed. A graphical concrete syntax seemed useful at first glance, but a textual concrete syntax turned out to be more effective, as we have already experienced with the domain, policy, and linking models. Furthermore, support for the specification of the mapping is desirable for the developer. It should e.g. be feasible to automatically derive identity mappings from the domain model for objects that are modeled identically at subsequent layers. Similarly, replacement mappings could be derived from similar objects, but this is obviously a more complex issue. Also, an implementation of the refinement process is being developed. As the refinement rules are provided in a declarative way, we use the declarative model transformation language Query/View/-Transformation (QVT) Relations [178] for our implementation. For an implementation of the refinement process with an imperative model transformation language such as QVT Operational Mappings [179], the imperative execution semantics of the refinement rules can be used as a starting point. Currently, the refinement process overwrites any previously refined policies at lower layers and re-generates them completely. An incremental refinement would be even more effective, i.e. a refinement that preserves manually adapted policies at lower layers and only generates the missing ones. Last but not least, the declarative definition of the refinement rules allows to extend our approach by reverse engineering. By inverting the implications within the refinement rules, the top-down refinement process can be turned into a bottom-up abstraction process that transforms low-level technical models into high-level functional ones. This is particularly interesting if changes at a lower level of abstraction should be reflected at a higher level.

5

Case Studies

To demonstrate the applicability of our policy specification and refinement approach, we present two case studies in sections 5.1 and 5.2. Both of them represent real-world scenarios that we developed in joint research projects with industry. Finally, section 5.3 provides a conclusion and an outlook.

The first case study deals with the management of Next Generation Mobile Networks (NGMN). For this purpose, a global telecommunications company provided various use cases from their operational business. In collaboration with this company, we developed solutions to automate configuration, performance, and fault management processes in mobile networks. We use Event Condition Action (ECA) policies to specify the decision logic, i.e. to specify which processes are executed under which circumstances. At the same time, our solutions provide control over the network to the human operator, who can change the respective policies at any time. Parts of our developments were filed as a patent application [49]. Last but not least, we developed an experimental system to demonstrate our solutions.

The second case study deals with bonus payments for employees. For this purpose, a medium-sized German consulting company provided a detailed insight into their business processes. Based on that, we developed solutions that automate the calculation of bonus payments. These solutions represent an automated business process that is supposed to replace the existing manual business process. We use ECA policies to specify the essential decision logic within the process. The decision logic can be changed at any time by changing the respective policies. Finally, we developed a prototype implementation of the automated business process.

5.1 Management of Mobile Networks

Future mobile networks will experience a continuous growth with regard to the number of Network Elements (NEs) and an increasing complexity with regard to their interrelations. A related trend is the seamless integration of multiple radio technologies into a single heterogeneous network. Both developments increase network management complexity and require new management concepts with a very high degree of automation, as currently discussed in the network research, operator, and standardization communities. In this context, Policy-Based Management (PBM) plays an important role as policies allow for automated network management processes. The usage of policies for mobile networks was recently considered in [19,43,45,47,50,109,118,180]. In order to address increasing management complexity in mobile networks, we developed policy-based solutions to automated decision making, and we present our solutions in this section.

5.1.1 Network Management

Management of a mobile network is a complex task as mobile networks have very high requirements on availability and stability. High expertise is required due to the distributed architecture of the underlying cellular network. Complexity arises from the high number of NEs to be deployed and managed and from interdependencies between their configurations. Besides that, mobile networks are usually very heterogeneous due to hardware from different vendors. The variety of deployed technologies and their proprietary operational paradigms are difficult to handle.

Operation, Administration, and Maintenance (OAM) of a mobile network is usually based on a centralized information system and organized in different management domains. Configuration Management (CM) deals with a consistent configuration of all network elements, Performance Management (PM) analyzes the efficiency of the current network configuration and seeks a more efficient one, and Fault Management (FM) detects and resolves errors that occur in the network. Any management domain focuses on different aspects and has a special view of the network and the configurations of the NEs. Network management is typically performed at an Operations and Maintenance Center (OMC) by human operators, who use their operational experience to find optimized configurations. Typically, management tasks are semi-automated only and tightly supervised. For this purpose, the network information system provides support by a set of planning and optimization tools. Nevertheless, a lot of human interaction is necessary, and manual control is time-consuming, expensive, and error-prone.

Long Term Evolution (LTE) is the standard for NGMN and was developed by the 3rd Generation Partnership Project (3GPP) [181]. Requests for reduction of Operational Expenditures (OPEXs) and for assuring high service quality call for changes in the way these networks are managed [182, 183]. In order to provide high-quality services in the future, management task automation is indispensable. Automation must go far beyond the possibilities that are used today such as scripting techniques. Although a lot of management tasks are similar, their automation is difficult as fulfilling the same task in a heterogeneous network can require different actions on different NEs. 3GPP has addressed this with the standardization of a framework for Self-Organizing Networks (SONs) [184]. SON features reflect network management knowledge and operator experience and are described in [32, 89, 185–187]. The introduction of SON features aims at reducing the workload on the operation and maintenance staff in order to free them from time-consuming and standard tasks, so that they are able to focus on more sophisticated problems. Even more important, a higher level of availability is gained by the automation of OAM tasks.

The objective is to gradually move towards a pure monitoring and guiding of the SON. Vendors are able to implement SON functions and integrate them one after another according to NGMN requirements. These SON functions realize different NGMN SON use cases and automatically perform the relevant CM actions. In a pure SON, only high-level guidelines for the network behavior need to be managed, instead of monitoring PM measurements and supervising CM actions manually.

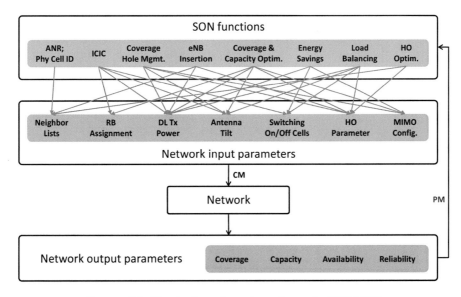

Figure 5.1: Control engineering view of a SON [47]

Figure 5.1 shows the control engineering view of a SON [47]. The overall state of the network is characterized by its output parameters such as radio access capacity and coverage. PM conveys measurements in the form of Key Performance Indicators (KPIs) to the controllers, which are in fact the SON functions such as Coverage and Capacity Optimization (CCO). These functions manipulate rather few network input parameters such as the Radio Transmission Power (TXP) and Remote Electrical Tilt (RET) of the antenna. These parameters in turn influence more than one network output parameter at the same time. Thus, there are potentially many dependencies between the SON functions, as can be seen from the figure. Therefore, the desired action from the perspective of an individual SON function needs to be coordinated with other potential actions from other functions, before the desired action is committed via the CM to the network.

5.1.2 Policy-Based Coordination

A SON usually contains a large number of independently acting SON functions across various use cases from all areas of network management [188]. Due to the introduction of self-organizing features into mobile networks, a potentially large number of SON functions are available. There will be SON functions for many aspects of Fault, Configuration, Accounting, Performance, and Security (FCAPS) management.

These functions are executed based on monitored network behavior, and this may lead to several functions being concurrently active in the same network area. Hence, many of these functions will have an impact on each other [90]. Uncoordinated execution of different functions with contradicting goals may lead to oscillating CM parameter changes and service degradation in the worst case. Furthermore, it is obvious that e.g. fault recovery functions have to be prioritized over network optimization functions. Unwanted effects may also arise within the same SON function type. The risk for such effects increases with the number of deployed SON functions. In order to avoid errors, oscillations, or even deadlocks in the configuration, coordination mechanisms are necessary. SON function coordination is indispensable in order to align the executed SON functions and thus to assure that they achieve the desired effect. While individual SON functions are already addressed in research projects such as SOCRATES [88, 89], only little attention has been paid to their coordination [189, 190].

Coordination Mechanism

On this account, we developed a coordination mechanism that uses policy-based decision making [45, 47]. First, we analyzed the SON functions to detect the potential interactions and conflicts between their actions. As a result of this analysis, we derived restrictions on the function execution to ensure that only non-conflicting SON functions are executed simultaneously. Whenever the execution of a SON function is requested, the previously and currently executed functions are checked for conflicts with the requested function. Based on this comparison, a decision is made. This decision can be to acknowledge (execute immediately), reschedule (execute a later point in time), or not acknowledge (not execute at all) the requested function. In addition, previously executed functions can be rolled back (recovering the state before their execution) or currently executed functions can be terminated (aborting their execution). Thus, we coordinate the SON function execution in a preventive way that avoids conflicts before they arise at all. We finally specified the decision logic with a set of ECA policies that automatically enforce the respective decisions. Still, the network is under control of the human operator, who can change these policies at any time. Some example policies are shown in section 5.1.3. An architecture where our policy-based coordination mechanism can be embedded is shown in [191].

Impact Area and Time

For purposes of coordination, we had to consider spatial and temporal dependencies between the NEs.

- Actions of a SON function on a single NE may have impact on adjacent NEs. Other actions on these NEs may result in a conflict and thus in an instable and undesired state of the network. NEs where a SON function possibly has impact are called the impact area of this function. The dependency does not necessarily relate to the same configuration parameter, but can also cover different parameters. It can even relate to physical network characteristics such as cell coverage.

- Actions of a SON function on an NE also have a temporal characteristic, i.e. actions have impact only within a limited time interval. This interval is called the impact time of the SON function. The impact time of a SON function typically exceeds the execution time of the function as it considers the actual impact on a measurable physical characteristic of the system, which may take a longer time.

Experimental System

An evaluation of the coordination mechanism is important in order to prove its applicability. As this is not possible with an operational network, we developed an experimental system that enables the simulation of real-world

coordination cases [46, 48]. The experimental system represents a technical solution and a proof of concept of our solutions. It consists of different
modules that are connected with each other by a message bus. The essential
modules are shown in figure 5.2.

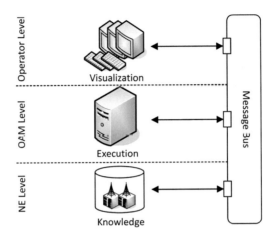

Figure 5.2: Architecture of the experimental system [45]

The Visualization module at the operator level represents a management
application that displays the current network configuration graphically on
a map. It allows to visualize complex scenarios such as the simultaneous execution of different SON functions. At the OAM level, the Execution module realizes the coordination mechanism. It contains the Ponder2 policy
engine [102] to enforce the coordination policies. The Knowledge module
at the NE level represents an information store for NEs and their configurations. The modular architecture makes the experimental system highly
extensible. Other systems such as an existing alarm system can be integrated via message wrappers. This even allows to integrate real events and
the actual SON functions from an operational network.

5.1.3 Physical Cell ID Assignment

We analyzed various SON functions with respect to their interrelations in
order to determine how the functions must be coordinated with each other.
This includes coordination of different instances of the same SON function.
In this section, we present the SON function for Physical Cell Identifier (PCI)
assignment and explain the requirements on its coordination.

The PCI is a fundamental parameter within LTE radio networks [32]. It
is used as a regionally unique identifier on the physical layer for the cells
and plays an essential role for the network operation. Without properly as-

signed PCIs, no radio communication is possible. Cell identification during handovers is crucial and has to be performed reliable and very quickly. The PCI parameter is used for cell identification as it is a reference signal sequence on the physical layer that can be accessed within 5 ms. Cells could also be identified by their globally unique EUTRAN Global Cell ID (EGCI), but accessing this parameter takes much more time and is therefore omitted whenever possible.

Assignment Restrictions

On the physical layer, 168 different reference signal sequences are available, and each of them is combined with three pseudo orthogonal codes. This results in a maximum of 504 usable PCIs. Obviously, there are a lot more than 504 cells in an LTE network, so that PCIs must be reused [192]. Furthermore, the PCI range is often fragmented in order to handle different cell types, vendor-specific cells, or assignment at country or license borders. As a result, only a partly PCI range with less than 504 possibilities is available for each cell.

Besides that, the assignment of PCIs to neighboring cells is restricted. Two adjacent cells are called direct neighbors if they have common coverage. These cells are also called 1st degree neighbors. 2nd degree neighbors of a cell are all direct neighbors of the 1st degree neighbors of this cell, without the cell itself and without its 1st degree neighbors. In general, n-th degree neighbors of a cell are all direct neighbors of the (n-1)-th degree neighbors of this cell, without the cell itself and without its (n-1)-th degree neighbors. The following two restrictions are crucial for the PCI assignment.

- **Collision-free:** Any two direct neighbors must be assigned different PCIs. If two direct neighbors use the same PCI, this is called a collision, as shown in figure 5.3a. Collisions can lead to situations where mobile devices in the common coverage area loose their connection or are not able to detect either of the cells.

- **Confusion-free:** Any direct neighbors of a cell must be assigned different PCIs. A cell is called confused if two or more of its direct neighbors are assigned identical PCIs, as shown in figure 5.3b. Confusions primarily have an impact on handover handling. A cell stores possible handover candidates in a table with the PCI as identifier for the respective target cells. Handovers are requested by mobile devices, which report the PCI of the desired target cell. If a mobile device requests a handover to an unknown target cell, but the PCI of this target cell is already used for another neighbor cell in the table, the handover will fail as it is prepared with the wrong cell.

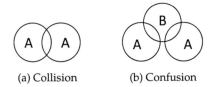

(a) Collision   (b) Confusion

Figure 5.3: Assignment restrictions

Assignment Scenarios

The following two scenarios must be taken into account for the PCI assignment.

- **Initial setup:** This is a scenario where all cells of a network, subnetwork, or large cluster of cells are not active until all of them have received a valid configuration. For two reasons, this is a rather uncritical scenario. First, the cell configurations can be determined in advance with knowledge about all cells. Second, all cell configurations can be applied at once without any reconfigurations of active cells.

- **Evolutionary growth:** In this scenario, the topology of the already deployed and operating network is changed, e.g. by adding cells or changing cell configurations. Evolutionary growth is critical as it requires reconfigurations of active cells in some scenarios. For example, confusion is inevitable if a new cell is added to the network and has a common coverage area with two cells that have identical PCIs, but have previously not been neighbors.

Assignment Strategies

As explained above, variations of radio propagation properties and changes of the network topology are important reasons for PCI reconfigurations at the cells of the network. Actually, these reconfigurations are rather frequent due to the dynamic nature of mobile networks. Each reconfiguration of a PCI causes a reboot of the base station and a service interruption within the cell. For this reason, it is important to provide an assignment that proactively reduces the number of reconfigurations.

The automated configuration of radio parameters is a key use case in NGMN [193]. For this purpose, we developed an automated assignment algorithm that calculates valid PCIs and covers both assignment scenarios [44, 194]. In order to carefully deal with PCIs as a limited resource, our algorithm minimizes the number of reconfigurations with the following two assignment strategies.

- **Maximum reuse:** This strategy uses as few different PCIs as possible and assigns the same PCIs repeatedly at cells with a minimum distance, as far as possible under the assignment restrictions. In order to minimize the number of reconfigurations, this strategy should be applied if cells are added primarily at the borders of coverage areas to extend the network coverage.

- **Maximum dispersion:** With this strategy, PCIs are assigned in a way that any two cells with the same PCI have a maximal distance from each other, as far as possible with the limited number of different PCIs. This strategy should be applied in cases where cells are added primarily within coverage areas to increase the network capacity.

To calculate a collision-free and confusion-free PCI, our algorithm takes into account the PCIs of the 1st and 2nd degree neighbors of the target cell. It is important that these PCIs remain stable during the calculation. For this reason, the algorithm uses the target cell itself and its 1st and 2nd degree neighbors as impact area. This impact area ensures that no two calculations are performed at the same cells simultaneously. The algorithm uses graph coloring [195] to determine an optimized PCI assignment. In comparison with other approaches [196, 197], it especially ensures that a valid PCI will be found and that no infinite reconfiguration loop will occur.

Safety Margin

Operators often have additional requirements on the PCI assignment. One example are cell outages. A cell outage can have multiple reasons and mostly results in a coverage hole within a certain area. An obvious method to compensate a coverage hole is to extend the coverage area of the surrounding cells. However, this results in new neighborships with a high risk of new collisions or confusions as cells with the same PCI might become neighbors. Another example is multipath propagation. Due to reflection from terrestrial objects, the same radio signal often reaches a mobile device by more than one path. Mobile devices usually select the strongest available signal and switch to a stronger signal of the same cell if one is detected. If a mobile device receives signals from different cells with the same PCI, it cannot distinguish the two cells. Switching to a signal from an actually different cell results in a connection drop. Especially in urban areas with a large number of small cells, a high reuse rate of PCIs is problematic as the conditions for radio wave propagation can change easily due to changing weather or construction. Figure 5.4 shows the actual coverage of cells in an urban area and illustrates different cells in different colors. In such a scenario, cells may become neighbors without configuration changes in the network, and this again implies a high risk of new collisions or confusions.

Figure 5.4: Actual coverage of cells in an urban area

Safety margins are a preventive means against collisions and confusions caused by coverage hole compensation or multipath propagation, and they reduce the probability of these issues. With a safety margin, cells are assigned PCIs that are not used by 3rd or higher degree neighbors, and this also implies a larger impact area. Simulations showed that enough different PCIs are still available for a valid assignment in realistic scenarios. Formally, a safety margin of n means that the PCI of a cell must not be reused by its n-th degree neighbors nor by neighbors of lower degree. With a safety margin of 2, the PCI assignment is at least free of collisions and confusions [44].

SON Function

In order to automate the PCI assignment, we realized a respective SON function. This SON function assigns a valid PCI to a cell with respect to the assignment restrictions (collision-free and confusion-free). The assignment of a PCI to a cell includes the calculation of a valid PCI and the enabling of this PCI at the cell. The SON function can also assign valid PCIs to a set of cells at once, and it supports PCI assignment in both scenarios (initial setup and evolutionary growth). The SON function is executed on request. A PCI can be requested in different situations, e.g. when a new cell is added to the network, the coverage of a cell changed, a new neighbor was added to a cell, or a PCI conflict was detected.

We use policies to control the execution of the SON function. Here, we described each situation that requests a PCI assignment with a respective event. Whenever such an event occurs, the respective policy triggers the execution of the SON function. The policy analyzes the context of the request in its condition, which can e.g. cover the time of the request and the safety margin of the affected cells. In its action, the policy triggers the PCI assignment or coordinates it with possibly conflicting SON functions. Such policies are also called trigger policies. The operator uses them to control and change the decision logic of the function execution.

Apart from automation, flexibility is an important aspect. Different operators may have different objectives for the PCI assignment in their networks. For this purpose, the SON function can be configured to apply a particular assignment strategy (maximum reuse or maximum dispersion) and a particular safety margin (2 or higher). Even within a network, it is not useful to have one fixed assignment strategy and safety margin. For this purpose, these settings can be changed at any time and then apply for any PCI assignment until they are changed again. They should be set to fit the context of different assignments. In order to ensure a robust network configuration, PCI assignments in urban regions with a high cell density should use a higher safety margin compared to assignments with a low density in rural areas.

We use policies to dynamically determine an appropriate assignment strategy and safety margin for each assignment. As described above, PCI assignments are requested by events. Whenever such an event occurs, the respective policy analyzes the context of the request in its condition, which can e.g. cover the reason for the assignment or the configuration of the respective cells. The policy then chooses appropriate settings in its action. Such policies are also called configuration policies. The operator uses them to control and change the decision logic of the configuration settings whenever necessary.

5.1.4 Model-Based Policy Specification

In this section, we show the specification of a trigger policy and a configuration policy at the operator level. At this level, we represent the policies from a management point of view that hides technical details from the operator. Later, we refine the specified policies into a technical representation at the OAM level. For this purpose, we first specify a domain model at the operator level, as shown in figure 5.5. The formal representation is provided in tables A.1 to A.6.

We describe the domain for PCI assignment with a domain model. We represent a cell in the mobile network with the concept *cell* and describe each cell with four properties: a unique *id*, the *location* and *radius* representing the coverage of the cell, and the *pci* of the cell. Furthermore, we represent a request for assigning a PCI at a particular cell with the concept *pciRequest*, and we indicate the respective cell with the relationship *targetCell*. We also model various operations: *getDensityOfArea* determines how dense the cells at a particular *dLocation* are located, *getPCIRange* determines the range of available PCIs, *setSafetyMargin* configures the *safetyMargin* for the following PCI assignments, *isAtDaytime* determines whether a particular *dTime* is at daytime, and *reschedule* reschedules a particular *rRequest*.

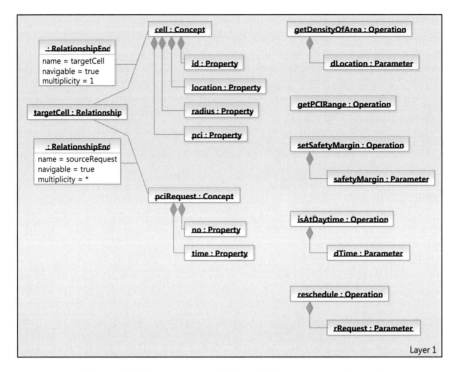

Figure 5.5: Domain model for PCI at operator level

Trigger Policy

We use the trigger policy to avoid PCI reconfigurations at busy times where a large number of mobile devices are connected. As mentioned before, each reconfiguration causes a service interruption within the cell, and any active voice or data connection with a mobile device will be dropped. For this reason, we desire to reschedule PCI reconfigurations that are requested in daytime to nighttime. A summary of the policy is provided in table 5.1. Figure 5.6 shows the respective policy model, and figure 5.7 shows the respective linking model. The formal representation is provided in tables A.17 to A.21 and A.27 to A.33.

Table 5.1: Summary of the trigger policy at operator level

Event:	A PCI assignment is requested
Condition:	The PCI request occurs at daytime
Action:	Reschedule the PCI assignment request

We represent any situation in which a PCI assignment is requested at a cell with the policy event *triggerEvent1*. We modeled such a situation with the generic concept *pciRequest* in the domain model. Therefore, we now link this concept to the event in the linking model.

In the policy condition *triggerCondition1*, we check the timing constraints. For this purpose, we invoke the operation *isAtDaytime* and pass the property *time* of the request to the parameter *dTime* of the operation. This property is visible in the policy event via the path *triggerEvent1 - pciRequest - time*. The boolean return value of the operation is used as value of the condition.

If the condition applies, we use the policy action *triggerAction1* to reschedule the assignment request. For this purpose, we invoke the operation *reschedule* and pass the property *no* to the parameter *rRequest* of the operation. This property represents the unique number of the request and is visible in the policy event via the path *triggerEvent1 - pciRequest - no*.

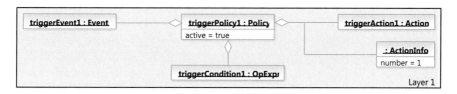

Figure 5.6: Policy model for the trigger policy at operator level

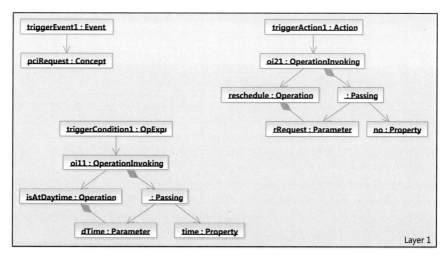

Figure 5.7: Linking model for the trigger policy at operator level

Configuration Policy

We use the configuration policy to configure the safety margin according to the particular request for the PCI assignment. The objective of the policy is to realize the maximum reuse strategy, i.e. to use as few different PCIs as possible in particular scenarios. The policy addresses two scenarios. Firstly, a maximum reuse makes sense for sparse cells, i.e. in areas with few cells only. This is typically the case in rural areas where reconfigurations are unlikely and multipath propagation is not an issue. Secondly, a maximum reuse should be applied if only few different PCIs are available at all. In both scenarios, it makes sense to apply the maximum reuse strategy by setting the safety margin to 2 as this is the minimal setting to avoid PCI collisions and confusions. A summary of the policy is provided in table 5.2. Figure 5.8 shows the respective policy model, and figure 5.9 shows the respective linking model. The formal representation is provided in tables A.17 to A.21 and A.27 to A.33.

Table 5.2: Summary of the configuration policy at operator level

Event:	A PCI assignment is requested
Condition:	The target cell of the PCI request is located in a sparse area or only a partial PCI range is available
Action:	Apply a safety margin of 2 for the PCI assignment

Again, we use the generic concept *pciRequest* in the domain model to represent situations in which a PCI assignment is requested at a cell. Therefore, we link this concept to the policy event *configurationEvent1* in the linking model.

In the policy condition *configurationCondition1*, we combine two subconditions as disjunction. The first subcondition *densityCondition1* addresses the density of the cells in the network and checks whether the cell is located in a sparse area. For this purpose, we invoke the operation *getDensityOfArea* and pass the property *location* to the parameter *dLocation* of the operation. This property represents the location of the target cell and is visible in the policy event via the path *configurationEvent1 - pciRequest - targetCell - cell - location*. We then check in the subcondition whether the return value of the operation is equal to the literal *sparse*. The second subcondition *rangeCondition1* addresses the available PCIs and checks whether only a partial PCI range is available. For this purpose, we invoke the operation *getPCIRange* and check whether the return value is equal to the literal *partial*.

If the condition is met, we use the policy action *configurationAction1* to set a particular safety margin. For this purpose, we invoke the operation *setSafetyMargin* and pass the literal *2* to the parameter *safetyMargin* of the operation.

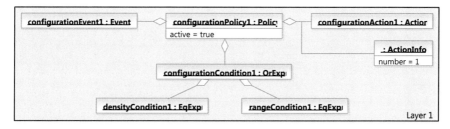

Figure 5.8: Policy model for the configuration policy at operator level

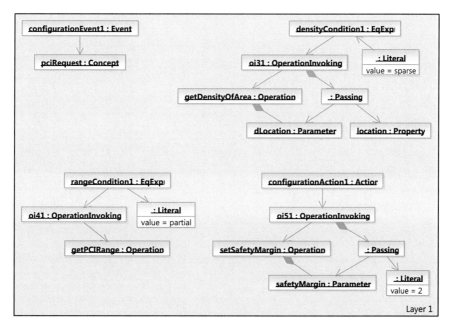

Figure 5.9: Linking model for the configuration policy at operator level

5.1.5 Model-Driven Policy Refinement

In order to enable a technical view, we specify a refined domain model at the OAM level, as shown in figure 5.10. We need this level to generate policies at the OAM level later. The formal representation of the refined domain model is provided in tables A.7 to A.12.

The concepts *cell* and *pciRequest* remain unchanged at this level. However, we now model four subconcepts of a PCI request that more precisely describe the reason for the assignment. The operation *getNumberOfNeighbors* determines the number of neighbor cells at a particular *nLocation*, and the operation *getNumberOfPCIs* determines how many different PCIs are available at all. The operation *setSafetyMargin* also remains unchanged.

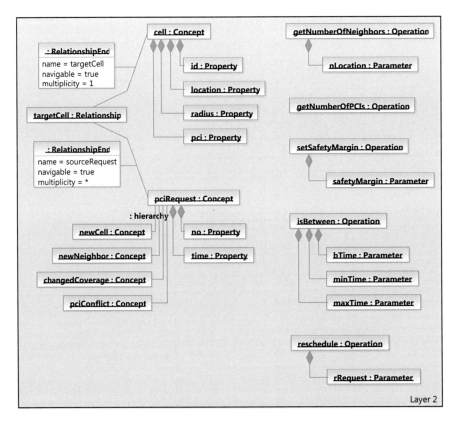

Figure 5.10: Domain model for PCI at OAM level

Then, we map the domain model from the operator to the OAM level and thus refine the management terms into technical ones. Figure 5.11 shows the respective mapping model. (5.1) to (5.13) and tables A.13 to A.16 show the formal representation of the mappings.

We refine the concept *pciRequest* with a split mapping into the four concepts *newCell*, *newNeighbor*, *changedCoverage*, and *pciConflict* (5.1). Thus, any occurrence of the concept *pciConflict* will be replaced with one out of the four refined concepts in refined policies. As we use a split mapping, the four possibilities will be proposed, and a decision will be requested. Furthermore, we use an identity mapping to leave the property *location* of a *cell* unchanged (5.2). Any occurrence of this property, e.g. when passed as argument, will remain the same in refined policies.

We use a replacement mapping to refine the operation *getDensityOfArea* with the literal *sparse* into the operation *getNumberOfNeighbors* with the literal 3 (5.3). This firstly means that any invoking of *getDensityOfArea* will be replaced in refined policies with an invoking of *getNumberOfNeighbors*. Secondly, whenever *getDensityOfArea* is used with the literal *sparse*, the literal 3 will be used with *getNumberOfNeighbors* instead. In other words, we refine a sparse area into a location with three neighbors. Furthermore, we refine the parameter *dLocation* into the parameter *nLocation* with a replacement mapping (5.4). This mapping makes sure that any argument passed to the parameter *dLocation* of the operation *getDensityOfArea* will be passed to the parameter *nLocation* of the refined operation *getNumberOfNeighbors*.

Furthermore, we use a replacement mapping to refine the operation *getPCIRange* with the literal *partial* into the operation *getNumberOfPCIs* with the literal 62 (5.5). Thus, any invoking of *getPCIRange* will be replaced in refined policies with an invoking of *getNumberOfPCIs*, and if the respective literals are used with the operation, they will be replaced accordingly. In this example, we refine a partial PCI range into a situation in which only 62 out of the total 504 PCIs are available.

We use two identity mappings to leave the operation *setSafetyMargin* with the literal 2 and its parameter *safetyMargin* unchanged (5.6,5.7). Any invoking of this operation will remain the same in refined policies.

We also refine the operation *isAtDaytime* into the operation *isBetween* with a replacement mapping (5.8). We refine the parameter *dTime* into the parameter *bTime* with another replacement mapping (5.9). Furthermore, we use two appearance mappings to assign the default literals *6am* and *12am* to the new parameters *minTime* and *maxTime* of the refined operation (5.10,5.11). We specify these literals with the mapping as they represent new information that cannot be derived. In this example, we refine daytime into a time interval that starts at 6am and end at 12am.

Finally, we use two identity mappings to leave the operation *reschedule* and its parameter *rRequest* unchanged (5.12,5.13). Any invoking of this operation will remain the same in refined policies.

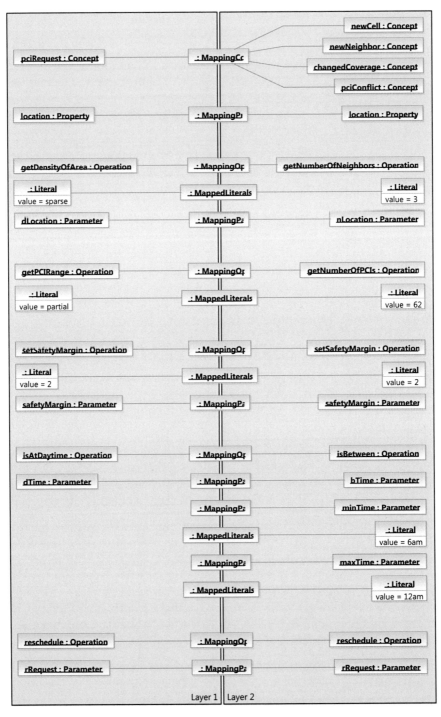

Figure 5.11: Mapping model for PCI

$$pciRequest \xrightarrow[1\to2]{Co} (newCell, newNeighbor,$$
$$changedCoverage, pciConflict) \tag{5.1}$$

$$location \xrightarrow[1\to2]{Pr} location \tag{5.2}$$

$$getDensityOfArea[sparse] \xrightarrow[1\to2]{Op} getNumberOfNeighbors[3] \tag{5.3}$$

$$dLocation \xrightarrow[1\to2]{Pa} nLocation \tag{5.4}$$

$$getPCIRange[partial] \xrightarrow[1\to2]{Op} getNumberOfPCIs[62] \tag{5.5}$$

$$setSafetyMargin[2] \xrightarrow[1\to2]{Op} setSafetyMargin[2] \tag{5.6}$$

$$safetyMargin \xrightarrow[1\to2]{Pa} safetyMargin \tag{5.7}$$

$$isAtDaytime \xrightarrow[1\to2]{Op} isBetween \tag{5.8}$$

$$dTime \xrightarrow[1\to2]{Pa} bTime \tag{5.9}$$

$$undef \xrightarrow[1\to2]{Pa} minTime[6am] \tag{5.10}$$

$$undef \xrightarrow[1\to2]{Pa} maxTime[12am] \tag{5.11}$$

$$reschedule \xrightarrow[1\to2]{Op} reschedule \tag{5.12}$$

$$rRequest \xrightarrow[1\to2]{Pa} rRequest \tag{5.13}$$

The refinement rules now imply the refined policies from the mapping model and the policy and linking models at the operator level. The single steps of the refinement were demonstrated in section 4.2.4. From an imperative point of view, the refinement rules apply the mapping model to the policy and linking models at the operator level, and they generate the policy and linking models at the OAM level. We explain the refinement of the policies by means of the two example policies in the following.

Trigger Policy

The event *triggerEvent1* of the policy is refined due to mapping (5.1). This mapping offers different possibilities for the refinement of the concept *pciRequest* to the operator, so the respective refinement rule requests a manual decision. In this example, we intend to handle the assignment of a PCI to a new neighboring cell, so we decide to use the refined concept *newNeighbor*. The respective refinement rule creates a refined event *configurationEvent2* and links it to the refined concept.

Mappings (5.8) to (5.11) are responsible for the refinement of the policy condition *triggerCondition1*. They map the operation *isAtDaytime* and the property *time* passed to the parameter *dTime* to the operation *isBetween* and the property *time* passed to the parameter *bTime*. Furthermore, they pass the default literals *6am* and *12am* to the parameters *minTime* and *maxTime*. The respective refinement rule creates a refined operation invoking. The refinement rule also creates a refined condition *triggerCondition2* and links it to the refined operation invoking.

The action *triggerAction1* of the policy is refined due to mappings (5.12) and (5.13). These leave the operation *reschedule* with the property *no* passed to the parameter *rNo* unchanged. The respective refinement rule creates a refined action *triggerAction2*, and the invoking of the operation remains the same in this action.

A summary of the policy is provided in table 5.3. Figure 5.12 shows the refined policy model, and figure 5.13 shows the refined linking model for the trigger policy. The formal representation is provided in tables A.22 to A.26 and A.34 to A.40.

Table 5.3: Summary of the trigger policy at operator level

Event:	A new neighbor is added
Condition:	The new neighbor is added between 6am and 12am
Action:	Reschedule the PCI assignment request

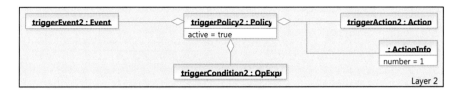

Figure 5.12: Policy model for the trigger policy at OAM level

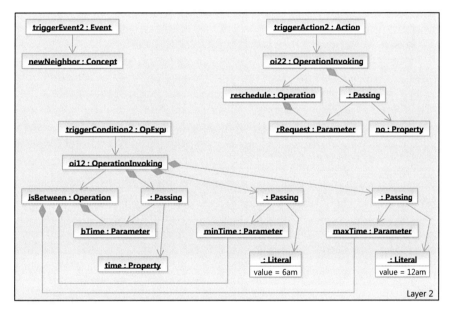

Figure 5.13: Linking model for the trigger policy at OAM level

Configuration Policy

The event *configurationEvent1* of the policy is refined due to mapping (5.1). Again, the respective refinement rule requests a manual decision due to the different possibilities for the refinement of the concept *pciRequest*. In this example, we intend to handle the assignment of a PCI to a new cell, so we decide to use the refined concept *newCell*. The respective refinement rule creates a refined event *configurationEvent2* and links it to the refined concept.

The policy condition *configurationCondition1* consists of two subconditions. Mappings (5.2) to (5.4) are responsible for the refinement of the first subcondition *densityCondition1*. They map the operation *getDensityOfArea* with the literal *sparse* and the property *location* passed to the parameter *dLocation* to the operation *getNumberOfNeighbors* with the literal *3* and the property *location* passed to the parameter *nLocation*. The second subcondition *rangeCondition1* is refined due to mapping (5.5), which maps the operation *getPCIRange* with its literal *partial* to the operation *getNumberOfNeighbors* with its literal *62*. The respective refinement rules create two refined subconditions *densityCondition2* and *rangeCondition2*. In these subconditions, the invokings of the operations are refined into invokings of the refined operations. The two subconditions are finally combined as disjunction into the refined condition *configurationCondition2*.

The action *configurationAction1* of the policy is refined due to mappings (5.6) and (5.7). They leave the operation *setSafetyMargin* with the literal *2* passed to the parameter *safetyMargin* unchanged. The respective refinement rule creates a refined action *configurationAction2*, and the invoking of the operation remains the same in this action.

A summary of the policy is provided in table 5.4. Figure 5.14 shows the refined policy model, and figure 5.15 shows the refined linking model for the configuration policy. The formal representation is provided in tables A.22 to A.26 and A.34 to A.40.

Table 5.4: Summary of the configuration policy at OAM level

Event:	A new cell is added
Condition:	The new cell has three neighbors or only 62 PCIs are available
Action:	Apply a safety margin of 2 for the PCI assignment

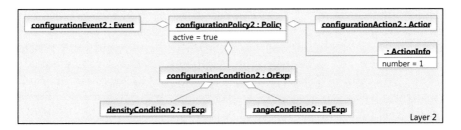

Figure 5.14: Policy model for the configuration policy at OAM level

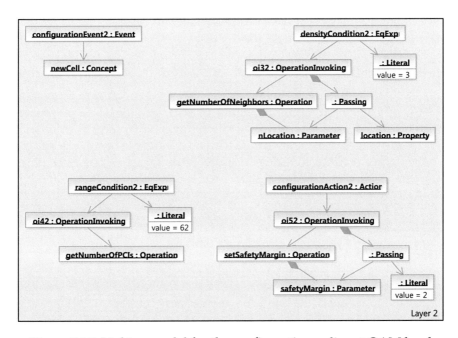

Figure 5.15: Linking model for the configuration policy at OAM level

5.2 Calculation of Bonus Payments

Since the industrial revolution, companies have experimented with compensation for employees. One aspect that has become more and more important is profit sharing. Profit sharing refers to direct or indirect payments to employees based on the profits of a company and was already mentioned in the 19th century [198]. At that time already, it was supposed to provide greater motivation to employees in order to work for the good of the enterprise and to promote closer feelings of partnership and shared purpose.

A related aspect are bonus payments. A bonus is a direct payment to an employee above the regular salary or hourly rate of pay. A lot of companies not only in the financial sector use bonus payments to their employees as an incentive to meet specific goals. While the focus of profit sharing is on the company as a whole, bonus payments focus on the performance of the individual employee. Usually, the individual goals of an employee are fixed at the beginning of a business year. At the end of the business year, the performance of the employee is measured depending on whether and to what extent the individual goals were achieved.

This case study focuses on bonus payments for consultants in a medium-sized German consulting company. In this company, the bonuses were calculated for each consultant using a manual business process. In order to replace the manual business process with an automated one, we developed a policy-based solution to the calculation of the bonuses. The respective requirements were specified in a master's thesis [199]. We then specified the decision logic within the business process with ECA policies.

5.2.1 Bonus Model

In the terminology of the consulting company, the bonus model describes the calculation logic of the bonus payments. Any consultant with a permanent contract of employment is eligible for a bonus payment. (5.14) to (5.17) show a simplified version of the bonus model for a consultant.

The individual performance of a consultant is represented as the number of performance points pp. Performance points are collected from the billable hours bh, the managed hours mh, and the acquired volume av. The billable hours represent the man hours of the consultant that are paid by customers; each man hour counts as one performance point. The managed hours represent the billable hours of subordinate consultants who are supervised by the consultant; each managed hour counts as 0.5 performance points. The acquired volume represents the contract value of projects acquired by the consultant; each 500 € of contract value counts as 1 performance point.

A bonus is only granted to a consultant if the individual goals are achieved or outperformed within the business year, i.e. if the performance points pp exceed the reference points rp. The reference points specify the individual goals that were fixed for the consultant at the begin of the business year. The difference between performance points and reference points is then multiplied with a certain point value pv. This value depends on the rank of the consultant and e.g. amounts to $9 €$ for a junior consultant and to $10 €$ for a senior consultant.

The extra bonus eb is granted as a percentaged supplement of the performance bonus pb above a certain threshold th. The percentage pc depends on the seniority of the consultant; the longer a consultant has been with the company, the higher it is. The total bonus tb is the sum of the performance bonus pb and the extra bonus eb.

$$pp = bh + \frac{mh}{2} + \frac{av}{500} \tag{5.14}$$

$$pb = \begin{cases} (pp - rp) \cdot pv, & pp \geq rp \\ 0, & \text{otherwise} \end{cases} \tag{5.15}$$

$$eb = \begin{cases} (pb - th) \cdot pc, & pb \geq th \\ 0, & \text{otherwise} \end{cases} \tag{5.16}$$

$$tb = pb + eb \tag{5.17}$$

5.2.2 Bonus Process

The bonus process is a realization of the bonus model as a business process. Whenever the bonus of a consultant should be calculated, this process is executed. For each consultant, the bonus process is executed at the end of the business year in order to determine the amount of the bonus payment. During the business year, each consultant can additionally request the intermediate bonus based on the individual performance until then. Each request triggers the bonus process once more.

Initially, the bonus process was a manual business process with three phases. Figure 5.16 shows the manual process in Business Process Model and Notation (BPMN). In the first phase, the relevant data for the calculation (billable hours, managed hours, and acquired volume) was gathered by the consultant and passed to the supervisor. In the second phase, the supervisor checked the reported data and passed it to the back office. If the data was considered incomplete or inconsistent, it was passed back to the consultant for completion or correction. Possibly, multiple iterations were necessary until the gathering of the data was finished. In the third phase, the staff in the back office performed a manual calculation of the bonus and sent a report to the consultant.

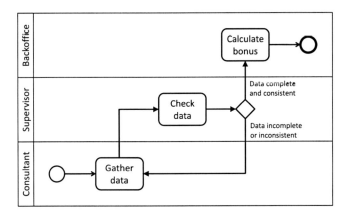

Figure 5.16: Manual bonus process

As the manual business process required a lot of human interaction, an automated solution was desired. For this purpose, we developed an automated business process for the accounting system. Figure 5.17 shows the resulting automated process in BPMN. In this process, we automated the gathering phase as far as possible. The relevant data is now gathered automatically from electronic documents that are created anyway, such as timekeeping or project order documents. Missing data can be entered in a web form. Thus, we minimized the manual effort for the consultant. As soon as all data is available, the process continues with the calculation. We were able to automate this phase completely. We realized the respective decision logic with ECA policies. Consequently, no more manual effort is necessary by the supervisor and the back office.

The focus in this case study is on the automated calculation phase. Figure 5.18 illustrates the logic within the bonus calculation. Although the calculation is specified in the bonus model, the bonus model leaves a certain degree of flexibility in every step. It e.g. does not decide about the point values for the calculation of the performance bonus or the percentages for the calculation of the extra bonus. These decisions are now made by ECA policies, and every step is controlled by a set of policies. Thus, the decision logic can be changed at any time by changing the respective policies.

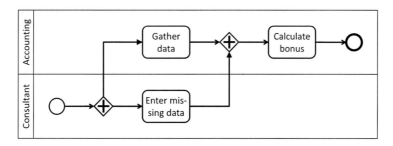

Figure 5.17: Automated bonus process

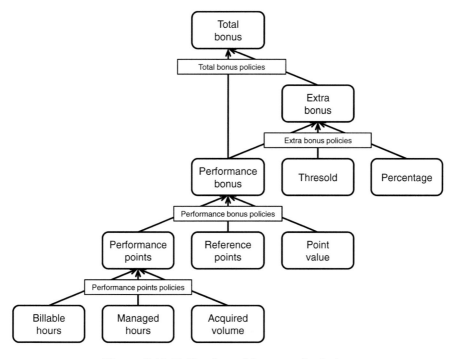

Figure 5.18: Policy-based bonus calculation

5.2.3 Model-Based Policy Specification

In this section, we show the specification of a performance bonus policy and an extra bonus policy at the application level. At this level, we represent the policies from an application point of view and thus hide technical details. Later, we refine the specified policies into a technical representation at the system level. For this purpose, we specify a domain model at the application level, as shown in figure 5.19 at first. The formal representation is provided in tables A.41 to A.46.

We describe the domain for bonus payments with a domain model. We represent a consultant with the concept *consultant* and describe it with three properties: a unique *name*, the *rank*, and the *seniority* of the consultant. We indicate with the concept *performancePointsCalculated* that the performance points of a consultant were calculated and are now available for further processing. This concept contains the respective *points* as property, and we reference the respective consultant with the relationship *cpp*. With the concept *performanceBonusCalculated*, we indicate that the performance bonus of a consultant was calculated. This concept contains the respective *bonus* as property, and we reference the respective consultant with the relationship *cpb*. Furthermore, we model two operations. The operation *calculatePerformanceBonus* calculates a performance bonus, taking the name of the respective consultant as parameter *pbConsultant* and the value per performance point as parameter *pointValue*. The operation *calculateExtraBonus* calculates an extra bonus, taking the name of the respective consultant and the percentage for the calculation as parameters *ebConsultant* and *percentage*.

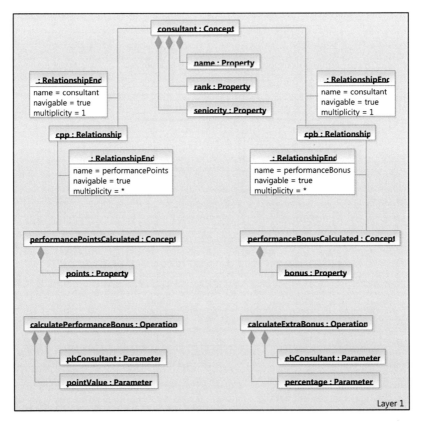

Figure 5.19: Domain model for bonus payments at application level

Performance Bonus Policy

We use performance bonus policies to proceed with the bonus calculation after the performance points of a consultant were calculated. These policies particularly decide about the point value to be used for the performance bonus calculation. We use different point values for different consultants according to the rank of the consultant. The performance bonus policy in table 5.5 is responsible for senior consultants. Although the reference points of a consultant are relevant for the performance bonus calculation as well, the policy does not deal with reference points as these are delivered by other policies. Figure 5.20 shows the respective policy model, and figure 5.21 shows the respective linking model. The formal representation is provided in tables A.57 to A.61 and A.67 to A.73.

Table 5.5: Summary of the performance bonus policy at application level

Event:	The performance points of a consultant are calculated
Condition:	The consultant is ranked senior consultant
Action:	Calculate the performance bonus of the consultant with a point value of $10 \in$

We represent any situation in which the performance points of a consultant were calculated with the policy event *performanceBonusEvent1*. We modeled this situation with the concept *performancePointsCaluculated* in the domain model. Therefore, we now link this concept to the event in the linking model.

In the policy condition *performanceBonusCondition1*, we compare the property *rank* of the consultant to the literal *seniorConsultant*. If both are equal, the condition is met. The property *rank* is visible in the policy event via the path *performanceBonusEvent1 - performancePointsCaluculated - cpp - consultant - rank*.

If the condition is met, we use the policy action *performanceBonusAction1* to calculate the performance bonus. For this purpose, we invoke the operation *calculatePerformanceBonus* and pass the property *name* of the consultant to the parameter *pbConsultant* and the literal *10* to the parameter *pointValue* of the operation. The property *name* is visible in the policy event via the path *performanceBonusEvent1 - performancePointsCaluculated - cpp - consultant - name*.

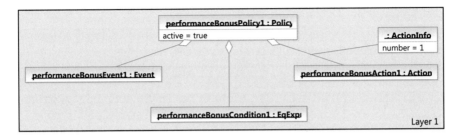

Figure 5.20: Policy model for the performance bonus policy at application level

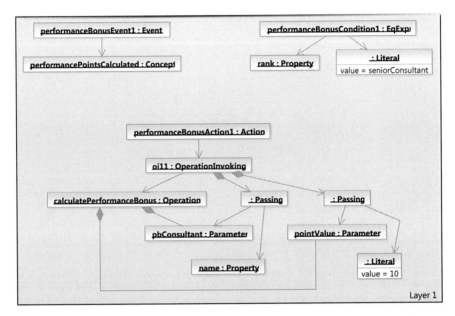

Figure 5.21: Linking model for the performance bonus policy at application level

Extra Bonus Policy

We use extra bonus policies to proceed with the bonus calculation after the performance bonus of a consultant was calculated. These policies particularly decide about the percentage to be used for the extra bonus calculation. We use different percentages for different consultants according to the seniority of the consultant. The extra bonus policy in table 5.6 is responsible for consultants who have been with the company for long time. Although the threshold is relevant for the extra bonus calculation as well, the policy does not deal with thresholds as these are delivered by other policies. Figure 5.22 shows the respective policy model, and figure 5.23 shows the respective linking model. The formal representation is provided in tables A.57 to A.61 and A.67 to A.73.

Table 5.6: Summary of the extra bonus policy at application level

Event:	The performance bonus of a consultant is calculated
Condition:	The consultant has been with the company for long time
Action:	Calculate the extra bonus of the consultant with a percentage of 40%

We represent any situation in which the performance bonus of a consultant was calculated with the policy event *extraBonusEvent1*. We modeled this situation with the concept *performanceBonusCaluculated* in the domain model. Therefore, we now link this concept to the event in the linking model.

In the policy condition *extraBonusCondition1*, we compare the property *seniority* of the consultant to the literal *long*. If both are equal, the condition is met. The property *seniority* is visible in the policy event via the path *extraBonusEvent1 - performanceBonusCaluculated - cpb - consultant - seniority*.

If the condition is met, we use the policy action *extraBonusAction1* to calculate the extra bonus. For this purpose, we invoke the operation *calculateExtraBonus* and pass the property *name* of the consultant to the parameter *ebConsultant* and the literal *40* to the parameter *percentage* of the operation. The property *name* is visible in the policy event via the path *extraBonusEvent1 - performanceBonusCaluculated - cpb - consultant - name*.

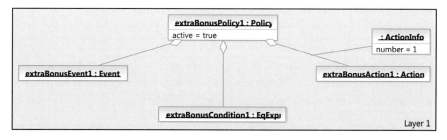

Figure 5.22: Policy model for the extra bonus policy at application level

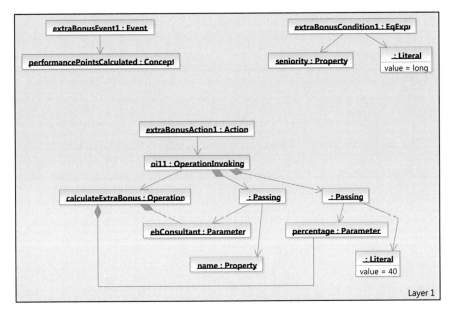

Figure 5.23: Linking model for the extra bonus policy at application level

5.2.4 Model-Driven Policy Refinement

In order to enable a technical view, we specify a refined domain model at the
system level, as shown in figure 5.24. We need this level in order to generate
policies at the system level later. The formal representation is provided in
tables A.47 to A.52.

The concepts *performancePointsCalculated* and *performanceBonusCalculated* remain unchanged at this level. We replace the property *rank* of concept *consultant* with the property *salaryGrade*, which provides a more detailed hierarchy of consultants with respect to their salary. The operations *calculatePerformanceBonus* and *calculateExtraBonus* also remain unchanged.

Then, we map the domain model from the application to the system level
and thus refine management terms into technical ones. Figure 5.25 shows
the respective mapping model. (5.18) to (5.27) and tables A.53 to A.56 show
the formal representation of the mappings.

We use two identity mappings to leave the concept *performancePointsCalculated* and the property *name* unchanged (5.18,5.19). Any occurrence of these
will remain the same in refined policies. Furthermore, we use a replacement mapping to refine the property *rank* with the literal *seniorConsultant*
into the property *salaryGrade* with the literal *5* (5.20). Any occurrence will

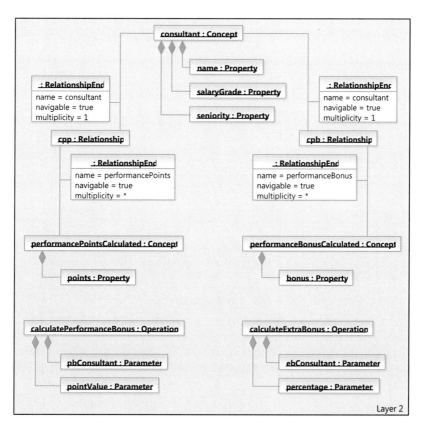

Figure 5.24: Domain model for bonus payments at system level

be replaced in refined policies accordingly. We use another identity mapping to leave the property *seniority* unchanged, but to refine its literal *long* into the literal *10* (5.21).

The operation *calculatePerformanceBonus* with the parameters *pbConsultant* and *pointValue* remains unchanged, as specified with the respective identity mappings (5.22,5.23,5.24). Also, the literal *10* of the parameter *pointValue* remains unchanged. Any invoking of this operation will remain the same in refined policies.

We use further identity mappings to leave the operation *calculateExtraBonus* with the parameters *ebConsultant* and *percentage* unchanged (5.25,5.26,5.27). The literal *40* of the parameter *pointValue* also remains unchanged. Any invoking of this operation will remain the same in refined policies.

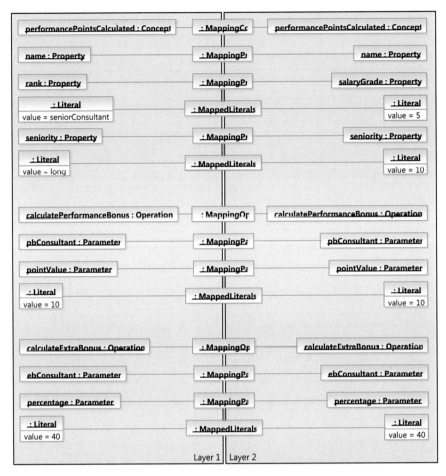

Figure 5.25: Mapping model for bonus payments

$$performancePointsCalculated \xrightarrow[1\to2]{Co} performancePointsCalculated \quad (5.18)$$

$$name \xrightarrow[1\to2]{Pr} name \quad (5.19)$$

$$rank[seniorConsultant] \xrightarrow[1\to2]{Pr} salaryGrade[5] \quad (5.20)$$

$$seniority[long] \xrightarrow[1\to2]{Pr} seniority[10] \quad (5.21)$$

$$calculatePerformanceBonus \xrightarrow[1\to2]{Op} calculatePerformanceBonus \tag{5.22}$$

$$pbConsultant \xrightarrow[1\to2]{Pa} pbConsultant \tag{5.23}$$

$$pointValue[10] \xrightarrow[1\to2]{Pa} pointValue[10] \tag{5.24}$$

$$calculateExtraBonus \xrightarrow[1\to2]{Op} calculateExtraBonus \tag{5.25}$$

$$ebConsultant \xrightarrow[1\to2]{Pa} ebConsultant \tag{5.26}$$

$$percentage[40] \xrightarrow[1\to2]{Pa} percentage[40] \tag{5.27}$$

The refinement rules now imply the refined policies from the mapping model and the policy and linking models at the application level. The single steps of the refinement were demonstrated in section 4.2.4. From an imperative point of view, the refinement rules apply the mapping model to the policy and linking models at the application level, and they generate the policy and linking models at the system level. We explain the refinement of the policies by means of the two example policies in the following.

Performance Bonus Policy

The event *performanceBonusEvent1* of the policy is refined due to mapping (5.18). It leaves the concept *performancePointsCalculated* unchanged. The respective refinement rule creates a refined event *performanceBonusEvent2* and links it to the refined concept.

Mapping (5.20) is responsible for the refinement of the policy condition *performanceBonusCondition1*. It maps the property *rank* with the literal *seniorConsultant* to the property *salaryGrade* with the literal *5*. The respective refinement rule creates a refined condition *performanceBonusCondition2* and links it to the refined property and literal.

The action *performanceBonusAction1* of the policy is refined due to mappings (5.19) and (5.22) to (5.24). They leave the operation *calculatePerformanceBonus* with the property *name* passed to the parameter *pbConsultant* and the literal *10* passed to the parameter *pointValue* unchanged. The respective refinement rule creates a refined action *performanceBonusAction2*, and the invoking of the operation remains the same in this action.

A summary of the policy is provided in table 5.7. Figure 5.26 shows the refined policy model, and figure 5.27 shows the refined linking model for the performance bonus policy. The formal representation is provided in tables A.62 to A.66 and A.74 to A.80.

Table 5.7: Summary of the performance bonus policy at system level

Event:	The performance points of a consultant are calculated
Condition:	The consultant is at salary grade 5
Action:	Calculate the performance bonus of the consultant with a point value of 10 €

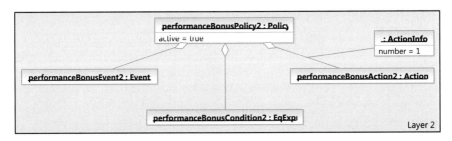

Figure 5.26: Policy model for the performance bonus policy at system level

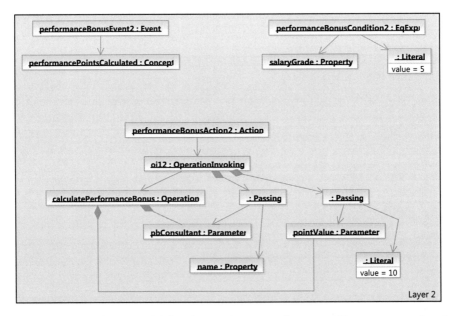

Figure 5.27: Linking model for the performance bonus policy at system level

Extra Bonus Policy

The event *extraBonusEvent1* of the policy is refined due to mapping (5.18). It leaves the concept *performanceBonusCalculated* unchanged. The respective refinement rule creates a refined event *extraBonusEvent2* and links it to the refined concept.

Mapping (5.21) is responsible for the refinement of the policy condition *extraBonusCondition1*. It leaves the property *seniority* unchanged, but maps the literal *long* to the literal *5*. The respective refinement rule creates a refined condition *extraBonusCondition2* and links it to the refined property and literal.

The action *extraBonusAction1* of the policy is refined due to mappings (5.19) and (5.25) to (5.27). They leave the operation *calculateExtraBonus* with the property *name* passed to the parameter *ebConsultant* and the literal *40* passed to the parameter *percentage* unchanged. The respective refinement rule creates a refined action *extraBonusAction2*, and the invoking of the operation remains the same in this action.

A summary of the policy is provided in table 5.8. Figure 5.28 shows the refined policy model, and figure 5.29 shows the refined linking model for the extra bonus policy. The formal representation is provided in tables A.62 to A.66 and A.74 to A.80.

Table 5.8: Summary of the extra bonus policy at system level

Event:	The performance bonus of a consultant is calculated
Condition:	The consultant has been with the company for at least 10 years
Action:	Calculate the extra bonus of the consultant with a percentage of 40%

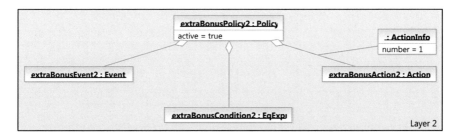

Figure 5.28: Policy model for the extra bonus policy at system level

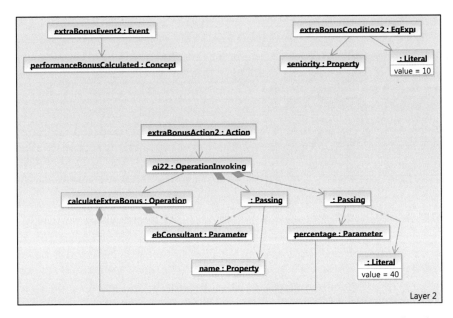

Figure 5.29: Linking model for the extra bonus policy at system level

5.3 Conclusion and Outlook

We presented two case studies as a proof of concept of the policy specification and refinement approach in this chapter. We conducted both case studies in collaboration with industry, and we showed that our approach enables the effective specification and refinement of ECA policies in realistic scenarios. These scenarios are reflected in different domains and different domain-specific policies. We specified concepts and policies for network management and bonus payments with the respective models. The refinement rules then generated machine-executable policies that can directly be mapped to the policy language of the underlying systems.

The case studies demonstrated the applicability of our approach with respect to the expressiveness of the models. The domain metamodel allowed us to describe the relevant parts of both domains at different levels of abstraction. The policy and linking metamodels offered enough means to describe the decision logic within the systems as domain-specific policies. We were able to express the refinement of both domains from a higher to a lower abstraction level with the mapping metamodel. The mapping patterns enabled powerful refinements, even in cases where the domain model only provided slight differences between the layers, as it was the case with the domain model for bonus payments.

Furthermore, the case studies demonstrated the automation of the refinement process. The refinement rules generated refined policies based on the provided models in an automated way. The refined policies provided the same functionality as the initial ones, but used technical concepts instead of abstract ones. Thus, they were machine-executable by the underlying systems and can directly be transformed into a policy language. However, the case study confirmed that a purely graphical syntax for the models is hardly applicable as graphical diagrams take a lot of space in more complex scenarios. As described before, an effective textual syntax is being developed.

The automated refinement also allowed us to change the behavior of the underlying network management and accounting systems by changing the high-level models. This was possible as the refined policies could be generated in an automated way, and this was possible even at runtime whenever a high-level model was changed. As changes of the abstract models were instantly reflected in their technical representation, even staff without programming skills managed to modify technical parameters of the system. This allowed to manage the behavior of the PCI assignment and the bonus calculation at a high level of abstraction and without considering technical details of the underlying systems. Such adaptability is especially useful as policy-based systems are very dynamic by nature and the respective policies are expected to undergo frequent changes.

6

Conclusion and Outlook

In this thesis, we presented an approach to the development of Event Condition Action (ECA) policies that makes use of techniques from Model-Driven Engineering (MDE) and raises the level of abstraction and automation in the development process. In section 6.1, we summarize our approach and emphasize the contributions with regard to the problems and challenges that we initially identified. In section 6.2, we discuss the contributions, and we provide an outlook on future work.

6.1 Summary

At the beginning of this thesis, we observed that Information and Communication Technology (ICT) systems have become more complex due to evolving hardware and software technology in recent years. Increasing complexity makes the development of these systems more difficult, and a lot of human effort is necessary for their operation. For this reason, we identified four problems, and for each problem, we derived a respective objective, presented an approach to achieve the objective, and developed a contribution that realizes the approach. Our contributions are illustrated in figure 6.1.

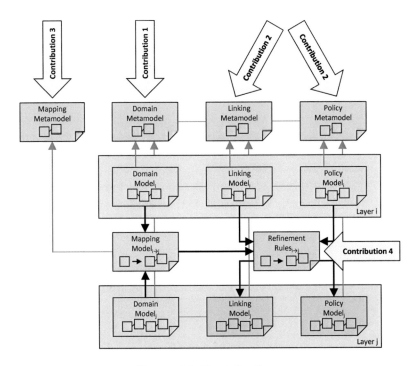

Figure 6.1: Contributions

First of all, different experts and expert groups use different terminology to describe a system and the underlying domain due to their different background, vocabulary, and expertise (*problem 1*). Instead, each expert group should agree on their individual understanding of the domain; for a better understanding between expert groups, all of them should use the same means to encapsulate domain knowledge (*objective 1*). For this purpose, we developed a generic domain metamodel. This metamodel allows to model the domain and domain-specific concepts at different abstraction layers. Each expert group is provided with a selected layer to represent their domain knowledge in their individual terminology. Views filter the parts of the domain that are relevant at a particular layer for a particular expert group (*approach and contribution 1*).

Secondly, complex systems are not only comprised of static components, but are often based on distributed and dynamic architectures, and this makes it difficult for developers to describe the behavior of the overall system (*problem 2*). Therefore, the behavior of a dynamic system should be described with decoupled rules to enable frequent adaption; each expert group should specify these rules with their individual focus, but with the same means (*objective 2*). To achieve this, we use policies that specify the reactive behavior of the system, i.e. how it should behave in particular situations. We developed a generic policy metamodel to specify policies at each abstraction layer. Furthermore, we developed a generic linking metamodel to use domain-specific content within the policies (*approach and contribution 2*).

Next, collaboration of application experts and technical experts in order to align high-level descriptions of system behavior with their low-level implementation is a time-consuming and error-prone task (*problem 3*). Therefore, higher-level descriptions of system behavior should be refined in an automated way for further processing; lower-level descriptions should be generated as input for the technical experts (*objective 3*). For this purpose, we establish cross-level mappings between domain-specific concepts. These mappings represent directions for the refinement of domain-specific policies, i.e. how to transform policies with domain-specific content from a higher layer to the subsequent lower layer. We developed a generic mapping metamodel for the modeling of the cross-layer mappings. This metamodel offers different mapping patterns to specify how the domain-specific concepts at a lower layer are derived from the precedent higher layer (*approach and contribution 3*).

Last but not least, systems in a dynamic environment are often expected to undergo frequent changes, and this demands for frequent adaption of system behavior at runtime and involves a lot of human effort (*problem 4*). Therefore, it should be possible to change the behavior of a system at runtime and in a flexible way, in particular without manually changing the implementation of the system (*objective 4*). To achieve this, we enable changes of system behavior by changing the respective policies in the model. We perform these changes in the model at the highest layer and propagate them down to the lowest layer. We developed refinement rules that apply the cross-layer mappings to the policy and linking models at runtime and generate refined models in an automated way. Finally, we proved that the refinement rules preserve the static semantics of the models, i.e. that valid models are refined into valid models again (*approach and contribution 4*).

6.2 Discussion

Our approach represents an effective progress to the development of domain-specific ECA policies. With techniques from MDE, we managed to raise the level of abstraction and automation in the development process. In particular, we provided the theoretical basis for an automated policy refinement. Although our approach helps to reduce accidental complexity in the development, it is not a silver bullet that eliminates essential development complexity [200]. The effectiveness of our approach depends on the developer and the modeling as the input models determine the potential quality of the generated output models. For this reason, we now highlight important aspects of our contributions and identify remaining challenges. Furthermore, we propose improvements and outline important aspects to be addressed in future.

Model-Based Policy Specification

The metamodels represent an essential contribution of our approach as they define the abstract syntax of domain-specific policies. Their key point is a proper separation of domain and policy aspects. The explicit modeling of the domain might be a complex and time-consuming task, but this is a one-time effort and afterwards simplifies the modeling of policies and the usage of domain-specific content within these policies.

The domain metamodel allows to specify a domain model. We deliberately decided not to use ontologies as we do not make use of their additional features such as greater expressiveness and reasoning techniques. This decision helps to keep our approach clear and simple. The policy metamodel allows to specify the structure of ECA policies in a policy model. Other policy types are not yet integrated as we focused on a declarative description on system behavior, and the ECA paradigm is an appropriate means to do this. To demonstrate the expressiveness of the models, we provided a running example and we presented two case studies that were developed in joint research projects with industry. These case studies showed that our approach enables the effective specification of ECA policies in realistic scenarios.

The specification of more complex domains and policies can be realized by an extension of the metamodels. This e.g. applies to a multiple inheritance of domain-specific concepts and to other policy types such as goal policies. In case of pure extensions, existing models remain valid instances of the extended metamodels. However, the representation of existing models might change if extensions require changes in the structure of the respective relations. An automated transformation of the existing models is desirable in this case.

A relational algebra defines the static semantics of the policies and enables their validation. The validity of the models is a prerequisite for their refinement into a lower level of abstraction and their transformation into executable code. We used a rather simple definition of validity that targets each policy separately. To be valid, a policy must only use domain-specific information that it can access, i.e. information that is passed via its events and that is visible at the layer of the policy. We did not validate multiple policies at a time in order to detect inconsistencies. Conflict detection and resolution is an important issue in Policy-Based Management (PBM), but existing approaches can be reused for this purpose, as presented in section 2.2.4. The formal representation of models with a relational algebra furthermore enables their persistent storage in a relational database. A respective database schema can directly be derived from the formal representation, and the relations can be mapped to the tables of the database.

Another important aspect is a concrete syntax to effectively specify the models. In our examples, we used Unified Modeling Language (UML) class diagrams to visualize the models in a graphical way as diagrams. However, diagrams take a lot of space in more complex scenarios, and experience showed that a purely graphical concrete syntax might be confusing. For this reason, an effective textual syntax is being developed, including tool support in terms of an editor and a textual Domain-Specific Language (DSL). The editor also validates the specified models according to their static semantics.

Model-Driven Policy Refinement

Another contribution of our approach is the refinement of the models across a flexible number of abstraction layers. This allows to specify policies at a high abstraction layer initially and avoids their direct implementation at a technical layer. The number of abstraction layers primarily depends on the complexity of the domain and the system to manage. For a given domain, a higher number of layers might enable more detailed mappings and thus a more precise refinement. In other words, the semantic gaps between every two layers become smaller. On the other hand, more layers usually require more initial effort for the specification of the domain and the mappings.

In our approach, the refinement of policies is automated by refinement rules that apply the mappings to the policies at the highest layer and thus generate refined policies at lower layers. In the refinement, the degree of automation highly depends on the specified mappings as these are the essential input for the refinement rules. If all specified mappings are deterministic, the refinement is fully automated. Often, multiple possibilities exist for a refinement as the refinement depends on information that is not available or cannot be expressed at design time. We addressed this with non-deterministic mapping patterns that allow to specify the different possibilities and to request a manual decision when the refinement is performed at runtime.

The formal semantics of the refinement process are defined by the refinement rules, which apply mappings to policies. The expressiveness of the refinement is determined by the available mapping patterns. Our mapping patterns replace an object or a combination of objects with another object or combination of objects. As shown in the case studies, the patterns already cover a variety of scenarios. However, more complex dependencies are imaginable that cannot be expressed with the currently available mapping patterns. In this case, the mapping metamodel can be extended in order to enable a more expressive set of mappings and refinement rules. An effective concrete syntax for the mapping model is being developed. Again, the question was whether a textual or a graphical representation of the mappings would be more appropriate. A graphical one initially seemed intuitive, but a textual one turned out to be more effective, as it was the case with the domain, policy, and linking models. Tool support for the specification of the mappings is being realized at the moment. It should also be possible to automatically derive certain mappings from the domain model, as for example identity mappings between objects that are modeled identically at subsequent layers, or replacement mappings between objects that are modeled similarly.

Our refinement rules provide the formal execution semantics of the refinement process and thus the basis for a straightforward implementation. Additionally, we analyzed the dependencies between the refinement rules and provided imperative execution semantics. Now, the refinement process is being implemented. As the refinement represents a model transformation, we use the model transformation language Query/View/Transformation (QVT) [178, 179] in our implementation. An incremental refinement is an important aspect, so that manually adapted policies are preserved at lower layers, and the missing ones are added. This is not the case at the moment as previously refined policies are overwritten at lower layers and re-generated completely. Another interesting aspect is an automated adaption of low-level policies based on monitored system behavior. This could be achieved with the help of machine learning techniques that allow for recognizing behavioral patterns from monitored data. Last but not least, reverse engineering of refined models is desirable to reflect manual changes and automated adaptions of low-level policies in the models at higher layers. As the refinement rules are defined in a declarative way, they can be turned around by inverting the implications within the refinement rules. Thus, it should be feasible to invert the top-down refinement process into a bottom-up abstraction process.

After their refinement to the lowest layer, policies are represented in a machine-executable way. The domain model then represents the concepts of the underlying system components, and the policy and linking models represent policies to control these components independently of any policy language. Now, executable code in a particular language can be generated from the models. Basically, this is possible for any language that is able to express ECA policies as defined by the policy metamodel. The feasibility of code generation for the policy language PonderTalk has already been shown [42]. For this purpose, model transformations generate the policy code in a fully automated way. This first involves a model-to-model transformation, which transforms the policy and linking models into an intermediate model-based representation of the target language. Then, a model-to-text transformation generates executable policy code in the target language. Although the domain-specific development of policies from abstract models to executable code is a complex task that will never be fully automated, our approach represents a solution that considerably raises the level of automation and motivates future research.

A

Formal Specifications

A.1 Management of Mobile Networks

A.1.1 Domain Model

Operator Level

Table A.1: View *Concept₁*

Layer	Name
1	cell
1	pciRequest

Table A.2: View *ConceptHierarchy₁*

Superconcept	Subconcept

Table A.3: View *Property₁*

Concept	Name
cell	id
cell	location
cell	radius
cell	pci
pciRequest	no
pciRequest	time

Table A.4: View *Operation₁*

Layer	Name
1	getDensityOfArea
1	getPCIRange
1	setSafetyMargin
1	isAtDaytime
1	reschedule

Table A.5: View *Parameter₁*

Operation	Name
getDensityOfArea	dLocation
setSafetyMargin	safetyMargin
isAtDaytime	dTime
reschedule	rNo

Table A.6: View *Relationship₁*

Concept1	Role1	Nav1	Mult1	Name	Mult2	Nav2	Role2	Concept2
pci-Request	source-Request	true	*	target-Cell	1	true	target-Cell	cell

OAM Level

Table A.7: View *Concept₂*

Layer	Name
2	cell
2	pciRequest
2	newCell
2	newNeighbor
2	changedCoverage
2	pciConflict

Table A.9: View *Property₂*

Concept	Name
cell	id
cell	location
cell	radius
cell	pci
pciRequest	no
pciRequest	time

Table A.8: View *ConceptHierarchy₂*

Superconcept	Subconcept
pciRequest	newCell
pciRequest	newNeighbor
pciRequest	changedCoverage
pciRequest	pciConflict

Table A.10: View *Operation₂*

Layer	Name
2	getNumberOfNeighbors
2	getNumberOfPCIs
2	setSafetyMargin
2	isBetween
2	reschedule

Table A.11: View *Parameter₂*

Operation	Name
getNumberOfNeighbors	nLocation
setSafetyMargin	safetyMargin
isBetween	bTime
isBetween	minTime
isBetween	maxTime
reschedule	rNo

Table A.12: View *Relationship₂*

Concept1	Role1	Nav1	Mult1	Name	Mult2	Nav2	Role2	Concept2
pci-Request	source-Request	true	*	target-Cell	1	true	target-Cell	cell

A.1.2 Mapping Model

Table A.13: Relation *MappingCo*

SourceLayer	SourceConcept	TargetLayer	TargetConcept	MappedLiterals
1	pciRequest	2	newCell	∅
1	pciRequest	2	newNeighbor	∅
1	pciRequest	2	changedCoverage	∅
1	pciRequest	2	pciConflict	∅

Table A.14: Relation *MappingPr*

SourceLayer	SourceProperty	TargetLayer	TargetProperty	MappedLiterals
1	location	2	location	∅

Table A.15: Relation *MappingOp*

SourceLayer	SourceOperation	TargetLayer	TargetOperation	MappedLiterals
1	getDensityOfArea	2	getNumberOfNeighbors	{(sparse,3)}
1	getPCIRange	2	getNumberOfPCIs	{(partial,62)}
1	setSafetyMargin	2	setSafetyMargin	{(2,2)}
1	isAtDaytime	2	isBetween	∅
1	reschedule	2	reschedule	∅

Table A.16: Relation *MappingPa*

SourceLayer	SourceParameter	TargetLayer	TargetParameter	MappedLiterals
1	dLocation	2	nLocation	∅
1	safetyMargin	2	safetyMargin	∅
1	dTime	2	bTime	∅
1	undef	2	minTime	{(undef,6am)}
1	undef	2	maxTime	{(undef,12am)}
1	rNo	2	rNo	∅

A.1.3 Policy Model

Operator Level

Table A.17: View *Policy₁*

Layer	Name	Active
1	triggerPolicy1	true
1	configurationPolicy1	true

Table A.18: View *Event₁*

Policy	Name
triggerPolicy1	triggerEvent1
configurationPolicy1	configurationEvent1

Table A.19: View *Condition₁*

Policy	Name	Type
triggerPolicy1	triggerCondition1	op
configurationPolicy1	configurationCondition1	or
undef	densityCondition1	eq
undef	rangeCondition1	eq

Table A.20: View *ConditionNesting₁*

OuterCondition	InnerCondition
configurationCondition1	densityCondition1
configurationCondition1	rangeCondition1

Table A.21: View *Action₁*

Policy	Name	No
triggerPolicy1	triggerAction1	1
configurationPolicy1	configurationAction1	1

OAM Level

Table A.22: View *Policy$_2$*

Layer	Name	Active
2	triggerPolicy2	true
2	configurationPolicy2	true

Table A.23: View *Event$_2$*

Policy	Name
triggerPolicy2	triggerEvent2
configurationPolicy2	configurationEvent2

Table A.24: View *Condition$_2$*

Policy	Name	Type
triggerPolicy2	triggerCondition2	op
configurationPolicy2	configurationCondition2	or
undef	densityCondition2	eq
undef	rangeCondition2	eq

Table A.25: View *ConditionNesting$_2$*

OuterCondition	InnerCondition
configurationCondition2	densityCondition2
configurationCondition2	rangeCondition2

Table A.26: View *Action$_2$*

Policy	Name	No
triggerPolicy2	triggerAction2	1
configurationPolicy2	configurationAction2	1

A.1.4 Linking Model

Operator Level

Table A.27: View EL_1

Event	Concept
triggerEvent1	pciRequest
configurationEvent1	pciRequest

Table A.28: View AL_1

Action	OpInv
triggerAction1	oi21
configurationAction1	oi51

Table A.29: View BEL_1

BinExpr	Arg1	Arg2
densityCondition1	oi31	sparse
rangeCondition1	oi41	partial

Table A.30: View OEL_1

OpExpr	OpInv
triggerCondition1	oi11

Table A.31: View $OIL1_1$

OpInv
oi11
oi21
oi31
oi41
oi51

Table A.32: View $OIL2_1$

OpInv	Operation
oi11	isAtDaytime
oi21	reschedule
oi31	getDensityOfArea
oi41	getPCIRange
oi51	setSafetyMargin

Table A.33: View $OIL3_1$

OpInv	Parameter	Argument
oi11	dTime	time
oi21	rNo	no
oi31	dLocation	location
oi51	safetyMargin	2

OAM Level

Table A.34: View EL_2

Event	Concept
triggerEvent2	newNeighbor
configurationEvent1	newCell

Table A.35: View AL_2

Action	OpInv
triggerAction2	oi22
configurationAction2	oi52

Table A.36: View BEL_2

BinExpr	Arg1	Arg2
densityCondition2	oi32	3
rangeCondition2	oi12	62

Table A.37: View OEL_2

OpExpr	OpInv
triggerCondition2	oi12

Table A.38: View $OIL1_2$

OpInv
oi12
oi22
oi32
oi42
oi52

Table A.39: View $OIL2_2$

OpInv	Operation
oi12	isBetween
oi22	reschedule
oi32	getNumberOfNeighbors
oi42	getNumberOfPCIs
oi52	setSafetyMargin

Table A.40: View $OIL3_2$

OpInv	Parameter	Argument
oi12	bTime	time
oi12	minTime	6am
oi12	maxTime	12am
oi22	rNo	no
oi32	nLocation	location
oi52	safetyMargin	2

A.2 Calculation of Bonus Payments

A.2.1 Domain Model

Application Level

Table A.41: View *Concept₁*

Layer	Name
1	consultant
1	performancePointsCalculated
1	performanceBonusCalculated

Table A.43: View *Property₁*

Concept	Name
consultant	name
consultant	rank
consultant	seniority
performancePointsCalculated	points
performanceBonusCalculated	bonus

Table A.42: View *ConceptHierarchy₁*

Superconcept	Subconcept

Table A.44: View *Operation₁*

Layer	Name
1	calculatePerformanceBonus
1	calculateExtraBonus

Table A.45: View *Parameter₁*

Operation	Name
calculatePerformanceBonus	pbConsultant
calculatePerformanceBonus	pointValue
calculateExtraBonus	ebConsultant
calculateExtraBonus	percentage

Table A.46: View *Relationship₁*

Concept1	Role1	Nav1	Mult1	Name	Mult2	Nav2	Role2	Concept2
consultant	consultant	true	1	cpp	*	true	performancePoints	performancePointsCalculated
consultant	consultant	true	1	cpb	*	true	performanceBonus	performanceBonusCalculated

System Level

Table A.47: View *Concept₂*

Layer	Name
2	consultant
2	performancePointsCalculated
2	performanceBonusCalculated

Table A.48: View *ConceptHierarchy₂*

Superconcept	Subconcept

Table A.49: View *Property₂*

Concept	Name
consultant	name
consultant	salaryGrade
consultant	seniority
performancePointsCalculated	points
performanceBonusCalculated	bonus

Table A.50: View *Operation₂*

Layer	Name
1	calculatePerformanceBonus
1	calculateExtraBonus

Table A.51: View *Parameter₂*

Operation	Name
calculatePerformanceBonus	pbConsultant
calculatePerformanceBonus	pointValue
calculateExtraBonus	ebConsultant
calculateExtraBonus	percentage

Table A.52: View *Relationship₂*

Concept1	Role1	Nav1	Mult1	Name	Mult2	Nav2	Role2	Concept2
consultant	consultant	true	1	cpp	*	true	performancePoints	performancePointsCalculated
consultant	consultant	true	1	cpb	*	true	performanceBonus	performanceBonusCalculated

A.2.2 Mapping Model

Table A.53: Relation *MappingCo*

SourceLayer	SourceConcept	TargetLayer	TargetConcept	MappedLiterals
1	performance-Points-Calculated	2	performance-Points-Calculated	∅

Table A.54: Relation *MappingPr*

SourceLayer	SourceProperty	TargetLayer	TargetProperty	MappedLiterals
1	name	2	name	∅
1	rank	2	salaryGrade	{(seniorConsultant,5)}
1	seniority	2	seniority	{(long,10)}

Table A.55: Relation *MappingOp*

SourceLayer	SourceOperation	TargetLayer	TargetOperation	MappedLiterals
1	calculate-Performance-Bonus	2	calculate-Performance-Bonus	∅
1	calculate-Extra-Bonus	2	calculate-Extra-Bonus	∅

Table A.56: Relation *MappingPa*

SourceLayer	SourceParameter	TargetLayer	TargetParameter	MappedLiterals
1	pbConsultant	2	pbConsultant	∅
1	pointValue	2	pointValue	{(10,10)}
1	ebConsultant	2	ebConsultant	∅
1	percentage	2	percentage	{(40,40)}

A.2.3 Policy Model

Application Level

Table A.57: View *Policy*$_1$

Layer	Name		Active
1	performanceBonusPolicy1		true
1	extraBonusPolicy1		true

Table A.58: View *Event*$_1$

Policy	Name
performanceBonusPolicy1	performanceBonusEvent1
extraBonusPolicy1	extraBonusEvent1

Table A.59: View *Condition*$_1$

Policy	Name	Type
performanceBonusPolicy1	performanceBonusCondition1	eq
extraBonusPolicy1	extraBonusCondition1	eq

Table A.60: View *ConditionNesting*$_1$

OuterCondition	InnerCondition

Table A.61: View *Action*$_1$

Policy	Name	No
performanceBonusPolicy1	performanceBonusAction1	1
extraBonusPolicy1	extraBonusAction1	1

System Level

Table A.62: View *Policy$_2$*

Layer	Name	Active
1	performanceBonusPolicy2	true
1	extraBonusPolicy2	true

Table A.63: View *Event$_2$*

Policy	Name
performanceBonusPolicy2	performanceBonusEvent2
extraBonusPolicy2	extraBonusEvent2

Table A.64: View *Condition$_2$*

Policy	Name	Type
performanceBonusPolicy2	performanceBonusCondition2	eq
extraBonusPolicy2	extraBonusCondition2	eq

Table A.65: View *ConditionNesting$_2$*

OuterCondition	InnerCondition

Table A.66: View *Action$_2$*

Policy	Name	No
performanceBonusPolicy2	performanceBonusAction2	1
extraBonusPolicy2	extraBonusAction2	1

A.2.4 Linking Model

Operator Level

Table A.67: View EL_1

Event	Concept
performanceBonus-Event1	performancePoints-Calculated
extraBonus-Event1	performancePoints-Calculated

Table A.68: View AL_1

Action	OpInv
performanceBonus-Action1	oi11
extraBonus-Action1	oi21

Table A.69: View BEL_1

BinExpr	Arg1	Arg2
performanceBonus-Condition1	rank	senior-Consultant
extraBonus-Condition1	seniority	long

Table A.70: View OEL_1

OpExpr	OpInv

Table A.71: View $OIL1_1$

OpInv
oi11
oi21

Table A.72: View $OIL2_1$

OpInv	Operation
oi11	calculatePerformanceBonus
oi21	calculateExtraBonus

Table A.73: View $OIL3_1$

OpInv	Parameter	Argument
oi11	pbConsultant	name
oi11	pointValue	10
oi21	ebConsultant	name
oi21	percentage	40

System Level

Table A.74: View EL_2

Event	Concept
performanceBonus-Event2	performancePoints-Calculated
extraBonus-Event2	performancePoints-Calculated

Table A.75: View AL_2

Action	OpInv
performanceBonus-Action1	oi12
extraBonus-Action1	oi22

Table A.76: View BEL_2

BinExpr	Arg1	Arg2
performanceBonus-Condition2	salaryGrade	5
extraBonus-Condition2	seniority	10

Table A.77: View OEL_2

OpExpr	OpInv

Table A.78: View $OIL1_2$

OpInv
oi12
oi22

Table A.79: View $OIL2_2$

OpInv	Operation
oi12	calculatePerformanceBonus
oi22	calculateExtraBonus

Table A.80: View $OIL3_2$

OpInv	Parameter	Argument
oi12	pbConsultant	name
oi12	pointValue	10
oi22	ebConsultant	name
oi22	percentage	40

B
Formal Proofs

B.1 Extension with a Policy

(3.69) holds for PM'_{la}

$\overset{(3.69)}{\Longleftrightarrow}$ $(la_1, po, act_1) \in Policy'_{la}$
$\wedge (la_2, po, act_2) \in Policy'_{la}$
$\Rightarrow la_1 = la_2 \wedge act_1 = act_2$

$\overset{(4.131)}{\Longleftrightarrow}$ $(la_1, po, act_1) \in Policy_{la} \cup (la, po', act')$
$\wedge (la_2, po, act_2) \in Policy_{la} \cup (la, po', act')$
$\Rightarrow la_1 = la_2 \wedge act_1 = act_2$

$\Longrightarrow$ $(la_1, po, act_1) \in Policy_{la} \wedge (la_2, po, act_2) \in Policy_{la}$
$\Rightarrow la_1 = la_2 \wedge act_1 - act_2$

$\overset{\text{I.H.}}{\Longrightarrow}$ $(la_1, po, act_1) \in Policy_{la+1} \wedge (la_2, po, act_2) \in Policy_{la+1}$
$\Rightarrow la_1 = la_2 \wedge act_1 = act_2$

$\overset{\substack{(4.100),(4.133),\\(4.73),(4.131),\text{I.H.}}}{\Longleftrightarrow}$ $((la_1, po, act_1) \in Policy_{la+1} \wedge (la_2, po, act_2) \in Policy_{la+1}$
$\Rightarrow la_1 = la_2 \wedge act_1 = act_2)$
$\wedge ((la_1, po, act_1) \in Policy_{la+1} \wedge (la_2, po, act_2) = (la + 1, po'', act')$
$\Rightarrow lu_1 = lu_2 \wedge ucl_1 = ucl_2)$
$\wedge ((la_2, po, act_2) \in Policy_{la+1} \wedge (la_1, po, act_1) = (la + 1, po'', act')$
$\Rightarrow la_1 = la_2 \wedge act_1 = act_2)$

$\Longleftrightarrow$ $((la_1, po, act_1) \in Policy_{la+1} \wedge (la_2, po, act_2) \in Policy_{la+1}$
$\Rightarrow la_1 = la_2 \wedge act_1 = act_2)$
$\wedge ((la_1, po, act_1) \in Policy_{la+1} \wedge (la_2, po, act_2) = (la + 1, po'', act')$
$\Rightarrow la_1 = la_2 \wedge act_1 = act_2)$
$\wedge ((la_1, po, act_1) = (la + 1, po'', act')) \wedge (la_2, po, act_2) \in Policy_{la+1}$
$\Rightarrow la_1 = la_2 \wedge act_1 = act_2)$
$\wedge ((la_1, po, act_1) = (la + 1, po'', act')) \wedge (la_2, po, act_2) = (la + 1, po'', act')$
$\Rightarrow la_1 = la_2 \wedge act_1 = act_2)$

$\Longleftrightarrow$ $((la_1, po, act_1) \in Policy_{la+1} \vee (la_1, po, act_1) = (la + 1, po'', act'))$
$\wedge ((la_2, po, act_2) \in Policy_{la+1} \vee (la_2, po, act_2) = (la + 1, po'', act'))$
$\Rightarrow la_1 = la_2 \wedge act_1 = act_2$

$\Longleftrightarrow$ $(la_1, po, act_1) \in Policy_{la+1} \cup (la + 1, po'', act')$
$\wedge (la_2, po, act_2) \in Policy_{la+1} \cup (la + 1, po'', act')$
$\Rightarrow la_1 = la_2 \wedge act_1 = act_2$

$\overset{(4.132)}{\Longleftrightarrow}$ $(la_1, po, act_1) \in Policy'_{la+1}$
$\wedge (la_2, po, act_2) \in Policy'_{la+1}$
$\Rightarrow la_1 = la_2 \wedge act_1 = act_2$

$\overset{(3.69)}{\Longleftrightarrow}$ (3.69) holds for PM'_{la+1}

B.2 Extension with an Event

$$(3.76) \text{ holds for } PM'_{la}$$

$$\overset{(3.76)}{\Longleftrightarrow} (po, ev) \in Event'_{la} \Rightarrow po \in Po \lor po = undef$$

$$\overset{(4.134)}{\Longleftrightarrow} (po, ev) \in Event_{la} \cup (po', ev') \Rightarrow po \in Po \lor po = undef$$

$$\Longrightarrow (po, ev) \in Event_{la} \Rightarrow po \in Po \lor po = undef$$

$$\overset{I.H.}{\Longrightarrow} (po, ev) \in Event_{la+1} \Rightarrow po \in Po \lor po = undef$$

$$\overset{(4.135),(4.71),(4.75)}{\Longleftrightarrow} ((po, ev) \in Event_{la+1} \Rightarrow po \in Po \lor po = undef)$$
$$\land po'' \in policiesOfLayer(la + 1)$$

$$\overset{(3.73),(3.70)}{\Longrightarrow} ((po, ev) \in Event_{la+1} \Rightarrow po \in Po \lor po = undef)$$
$$\land po'' \in Po$$

$$\Longleftrightarrow ((po, ev) \in Event_{la+1} \Rightarrow po \in Po \lor po = undef)$$
$$\land (po, ev) = (po'', ev'') \Rightarrow po \in Po \lor po = undef$$

$$\Longleftrightarrow (po, ev) \in Event_{la+1} \cup (po'', ev'') \Rightarrow po \in Po \lor po = undef$$

$$\overset{(4.136)}{\Longleftrightarrow} (po, ev) \in Event'_{la+1} \Rightarrow po \in Po \lor po = undef$$

$$\overset{(3.76)}{\Longleftrightarrow} (3.76) \text{ holds for } PM'_{la+1}$$

B.3 Extension with a Condition

$$(3.85) \text{ holds for } PM'_{la}$$

$$\overset{(3.85)}{\Longleftrightarrow} \quad (po, cd, ty) \in Condition'_{la} \Rightarrow po \in Po \lor po = undef$$

$$\overset{(4.138)}{\Longleftrightarrow} \quad (po, cd, ty) \in Condition_{la} \cup (po', cd', ty') \Rightarrow po \in Po \lor po = undef$$

$$\Longrightarrow \quad (po, cd, ty) \in Condition_{la} \Rightarrow po \in Po \lor po = undef$$

$$\overset{\text{I.H.}}{\Longrightarrow} \quad (po, cd, ty) \in Condition_{la+1} \Rightarrow po \in Po \lor po = undef$$

$$\overset{(4.139),(4.71),(4.75)}{\Longleftrightarrow} \quad ((po, cd, ty) \in Condition_{la+1} \Rightarrow po \in Po \lor po = undef)$$
$$\land po'' \in policiesOfLayer(la + 1)$$

$$\overset{(3.73),(3.70)}{\Longrightarrow} \quad ((po, cd, ly) \in Condition_{la+1} \rightarrow po \subset Po \lor po = undef)$$
$$\land po'' \in Po$$

$$\Longleftrightarrow \quad ((po, cd, ty) \in Condition_{la+1} \Rightarrow po \in Po \lor po = undef)$$
$$\land (po, cd, ty) = (po'', cd'', ty') \Rightarrow po \in Po \lor po = undef$$

$$\Longleftrightarrow \quad (po, cd, ty) \in Condition_{la+1} \cup (po'', cd'', ty') \Rightarrow po \in Po \lor po = undef$$

$$\overset{(4.140)}{\Longleftrightarrow} \quad ((po, cd, ty) \in Condition'_{la+1} \Rightarrow po \in Po \lor po = undef$$

$$\overset{(3.85)}{\Longleftrightarrow} \quad (3.85) \text{ holds for } PM'_{la+1}$$

$$(3.86) \text{ holds for } PM'_{la}$$

$$\stackrel{(3.86)}{\Longleftrightarrow} (po, cd_1, ty_1) \in Condition'_{la}$$
$$\wedge (po, cd_2, ty_2) \in Condition'_{la}$$
$$\Rightarrow cd_1 = cd_2 \wedge ty_1 = ty_2$$

$$\stackrel{(4.138)}{\Longleftrightarrow} (po, cd_1, ty_1) \in Condition_{la} \cup (po', cd', ty')$$
$$\wedge (po, cd_2, ty_2) \in Condition_{la} \cup (po', cd', ty')$$
$$\Rightarrow cd_1 = cd_2 \wedge ty_1 = ty_2$$

$$\Longrightarrow (po, cd_1, ty_1) \in Condition_{la} \wedge (po, cd_2, ty_2) \in Condition_{la}$$
$$\Rightarrow cd_1 = cd_2 \wedge ty_1 = ty_2$$

$$\stackrel{\text{I.H.}}{\Longrightarrow} (po, cd_1, ty_1) \in Condition_{la+1} \wedge (po, cd_2, ty_2) \in Condition_{la+1}$$
$$\Rightarrow cd_1 = cd_2 \wedge ty_1 = ty_2$$

$$\stackrel{\substack{(4.102),(4.139),(4.73),\\(4.138),\text{I.H.},(4.140)}}{\Longleftrightarrow} ((po, cd_1, ty_1) \in Condition_{la+1} \wedge (po, cd_2, ty_2) \in Condition_{la+1}$$
$$\Rightarrow cd_1 = cd_2 \wedge ty_1 = ty_2)$$
$$\wedge ((po, cd_1, ty_1) \in Condition_{la+1} \wedge (po, cd_2, ty_2) = (po'', cd'', ty')$$
$$\Rightarrow cd_1 = cd_2 \wedge ty_1 = ty_2)$$
$$\wedge ((po, cd_2, ty_2) \in Condition_{la+1} \wedge (po, cd_1, ty_1) = (po'', cd'', ty')$$
$$\Rightarrow cd_1 = cd_2 \wedge ty_1 = ty_2)$$

$$\Longleftrightarrow ((po, cd_1, ty_1) \in Condition_{la+1} \wedge (po, cd_2, ty_2) \in Condition_{la+1}$$
$$\Rightarrow cd_1 = cd_2 \wedge ty_1 = ty_2)$$
$$\wedge ((po, cd_1, ty_1) \in Condition_{la+1} \wedge (po, cd_2, ty_2) = (po'', cd'', ty')$$
$$\Rightarrow cd_1 = cd_2 \wedge ty_1 = ty_2)$$
$$\wedge ((po, cd_1, ty_1) = (po'', cd'', ty') \wedge (po, cd_2, ty_2) \in Condition_{la+1})$$
$$\Rightarrow cd_1 = cd_2 \wedge ty_1 = ty_2)$$
$$\wedge ((po, cd_1, ty_1) = (po'', cd'', ty') \wedge (po, cd_2, ty_2) = (po'', cd'', ty')$$
$$\Rightarrow cd_1 = cd_2 \wedge ty_1 = ty_2)$$

$$\Longleftrightarrow ((po, cd_1, ty_1) \in Condition_{la+1} \vee (po, cd_1, ty_1) = (po'', cd'', ty'))$$
$$\wedge ((po, cd_2, ty_2) \in Condition_{la+1} \vee (po, cd_2, ty_2) = (po'', cd'', ty'))$$
$$\Rightarrow cd_1 = cd_2 \wedge ty_1 = ty_2$$

$$\Longleftrightarrow (po, cd_1, ty_1) \in Condition_{la+1} \cup (po'', cd'', ty')$$
$$\wedge (po, cd_2, ty_2) \in Condition_{la+1} \cup (po'', cd'', ty')$$
$$\Rightarrow cd_1 = cd_2 \wedge ty_1 = ty_2$$

$$\stackrel{(4.140)}{\Longleftrightarrow} (po, cd_1, ty_1) \in Condition'_{la+1}$$
$$\wedge (po, cd_2, ty_2) \in Condition'_{la+1}$$
$$\Rightarrow cd_1 = cd_2 \wedge ty_1 = ty_2$$

$$\stackrel{(3.86)}{\Longleftrightarrow} (3.86) \text{ holds for } PM'_{la+1}$$

B.4 Extension with a Condition Nesting

(3.85) holds for PM'_{la}

$\stackrel{(3.85)}{\Longleftrightarrow}$ $(po, cd, ty) \in Condition'_{la} \Rightarrow po \in Po \vee po = undef$

$\stackrel{(4.143)}{\Longleftrightarrow}$ $(po, cd, ty) \in Condition_{la} \Rightarrow po \in Po \vee po = undef$

$\stackrel{\text{I.H.}}{\Longrightarrow}$ $(po, cd, ty) \in Condition_{la+1} \Rightarrow po \in Po \vee po = undef$

$\Longleftarrow$ $((po, cd, ty) \in Condition_{la+1} \Rightarrow po \in Po \vee po = undef)$
$\qquad \wedge ((po, cd, ty) = (undef, cd''_1, ty') \Rightarrow po \in Po \vee po = undef)$

$\Longleftarrow$ $(po, cd, ty) \in Condition_{la+1} \cup (undef, cd''_1, ty') \Rightarrow po \in Po \vee po = undef$

$\stackrel{(4.147)}{\longleftrightarrow}$ $(po, cd, ty) \in Condition'_{la+1} \rightarrow po \in Po \vee po = undef$

$\stackrel{(3.85)}{\Longleftrightarrow}$ (3.85) holds for PM'_{la+1}

$$(3.86) \text{ holds for } PM'_{la}$$

$$\overset{(3.86)}{\Longleftrightarrow} \quad (po, cd_1, ty_1) \in Condition'_{la}$$
$$\wedge (po, cd_2, ty_2) \in Condition'_{la}$$
$$\Rightarrow cd_1 = cd_2 \wedge ty_1 = ty_2$$

$$\overset{(4.143)}{\Longleftrightarrow} \quad (po, cd_1, ty_1) \in Condition_{la}$$
$$\wedge (po, cd_2, ty_2) \in Condition_{la}$$
$$\Rightarrow cd_1 = cd_2 \wedge ty_1 = ty_2$$

$$\overset{\text{I.H.}}{\Longrightarrow} \quad (po, cd_1, ty_1) \in Condition_{la+1} \wedge (po, cd_2, ty_2) \in Condition_{la+1}$$
$$\Rightarrow cd_1 = cd_2 \wedge ty_1 = ty_2$$

$$\overset{\substack{(4.102),(4.143),\\(4.145),\text{I.H.},(4.147)}}{\Longleftrightarrow} \quad ((po, cd_1, ty_1) \in Condition_{la+1} \wedge (po, cd_2, ty_2) \in Condition_{la+1}$$
$$\Rightarrow cd_1 = cd_2 \wedge ty_1 = ty_2)$$
$$\wedge ((po, cd_1, ty_1) \in Condition_{la+1} \wedge (po, cd_2, ty_2) = (undef, cd''_1, ty')$$
$$\Rightarrow cd_1 = cd_2 \wedge ty_1 = ty_2)$$
$$\wedge ((po, cd_2, ty_2) \in Condition_{la+1} \wedge (po, cd_1, ty_1) = (undef, cd''_1, ty')$$
$$\Rightarrow cd_1 = cd_2 \wedge ty_1 = ty_2)$$

$$\Longleftrightarrow \quad ((po, cd_1, ty_1) \in Condition_{la+1} \wedge (po, cd_2, ty_2) \in Condition_{la+1}$$
$$\Rightarrow cd_1 = cd_2 \wedge ty_1 = ty_2)$$
$$\wedge ((po, cd_1, ty_1) \in Condition_{la+1} \wedge (po, cd_2, ty_2) = (undef, cd''_1, ty')$$
$$\Rightarrow cd_1 = cd_2 \wedge ty_1 = ty_2)$$
$$\wedge ((po, cd_1, ty_1) = (undef, cd''_1, ty') \wedge (po, cd_2, ty_2) \in Condition_{la+1})$$
$$\Rightarrow cd_1 = cd_2 \wedge ty_1 = ty_2)$$
$$\wedge ((po, cd_1, ty_1) = (undef, cd''_1, ty') \wedge (po, cd_2, ty_2) = (undef, cd''_1, ty')$$
$$\Rightarrow cd_1 = cd_2 \wedge ty_1 = ty_2)$$

$$\Longleftrightarrow \quad ((po, cd_1, ty_1) \in Condition_{la+1} \vee (po, cd_1, ty_1) = (undef, cd''_1, ty'))$$
$$\wedge ((po, cd_2, ty_2) \in Condition_{la+1} \vee (po, cd_2, ty_2) = (undef, cd''_1, ty'))$$
$$\Rightarrow cd_1 = cd_2 \wedge ty_1 = ty_2$$

$$\Longleftrightarrow \quad (po, cd_1, ty_1) \in Condition_{la+1} \cup (undef, cd''_1, ty')$$
$$\wedge (po, cd_2, ty_2) \in Condition_{la+1} \cup (undef, cd''_1, ty')$$
$$\Rightarrow cd_1 = cd_2 \wedge ty_1 = ty_2$$

$$\overset{(4.147)}{\Longleftrightarrow} \quad (po, cd_1, ty_1) \in Condition'_{la+1}$$
$$\wedge (po, cd_2, ty_2) \in Condition'_{la+1}$$
$$\Rightarrow cd_1 = cd_2 \wedge ty_1 = ty_2$$

$$\overset{(3.86)}{\Longleftrightarrow} \quad (3.86) \text{ holds for } PM'_{la+1}$$

(3.96) holds for PM'_{la}

$\overset{(3.96)}{\Longleftrightarrow}$ $(cd_2, cd_1) \in ConditionNesting'_{la} \Rightarrow cd_2, cd_1 \in Cd$

$\overset{(4.142)}{\Longleftrightarrow}$ $(cd_2, cd_1) \in ConditionNesting_{la} \cup (cd'_2, cd'_1) \Rightarrow cd_2, cd_1 \in Cd$

$\Longrightarrow$ $(cd_2, cd_1) \in ConditionNesting_{la} \Rightarrow cd_2, cd_1 \in Cd$

$\overset{I.H.}{\Longrightarrow}$ $(cd_2, cd_1) \in ConditionNesting_{la+1} \Rightarrow cd_2, cd_1 \in Cd$

$\overset{(4.148),(4.83),(4.87)}{\Longleftrightarrow}$ $((cd_2, cd_1) \in ConditionNesting_{la+1} \Rightarrow cd_2, cd_1 \in Cd)$
$\wedge cd''_2 \in conditionsOfLayer(la+1)$
$\wedge cd''_1 \in conditionsOfLayer(la+1)$

$\overset{(3.93),(3.87)}{\longrightarrow}$ $((cd_2, cd_1) \in ConditionNesting_{la+1} \Rightarrow cd_2, cd_1 \in Cd)$
$\wedge cd''_2 \in Cd \wedge cd''_1 \in Cd$

$\Longleftrightarrow$ $((cd_2, cd_1) \in ConditionNesting_{la+1} \Rightarrow cd_2, cd_1 \in Cd)$
$\wedge (cd_2, cd_1) = (cd''_2, cd''_1) \Rightarrow cd_2, cd_1 \in Cd$

$\Longleftrightarrow$ $(cd_2, cd_1) \in ConditionNesting_{la+1} \cup (cd''_2, cd''_1) \Rightarrow cd_2, cd_1 \in Cd$

$\overset{(4.146)}{\Longleftrightarrow}$ $(cd_2, cd_1) \in ConditionNesting'_{la+1} \Rightarrow cd_2, cd_1 \in Cd$

$\overset{(3.96)}{\Longleftrightarrow}$ (3.96) holds for PM'_{la+1}

$$(3.97) \text{ holds for } PM'_{la}$$

$$\overset{(3.97)}{\Longleftrightarrow} \quad (cd_1, cd) \in ConditionNesting'_{la}$$
$$\wedge (cd_2, cd) \in ConditionNesting'_{la}$$
$$\Rightarrow cd_1 = cd_2$$

$$\overset{(4.142)}{\Longleftrightarrow} \quad (cd_1, cd) \in ConditionNesting_{la} \cup (cd'_2, cd'_1)$$
$$\wedge (cd_2, cd) \in ConditionNesting_{la} \cup (cd'_2, cd'_1)$$
$$\Rightarrow cd_1 = cd_2$$

$$\Longrightarrow \quad (cd_1, cd) \in ConditionNesting_{la} \wedge (cd_2, cd) \in ConditionNesting_{la}$$
$$\Rightarrow cd_1 = cd_2$$

$$\overset{\text{I.H.}}{\Longrightarrow} \quad (cd_1, cd) \in ConditionNesting_{la+1} \wedge (cd_2, cd) \in ConditionNesting_{la+1}$$
$$\Rightarrow cd_1 = cd_2$$

$$\overset{\substack{(4.103),(4.148),(4.85),\\(4.142),\text{I.H.},(4.146)}}{\Longleftrightarrow} \quad ((cd_1, cd) \in ConditionNesting_{la+1} \wedge (cd_2, cd) \in ConditionNesting_{la+1}$$
$$\Rightarrow cd_1 = cd_2)$$
$$\wedge ((cd_1, cd) \in ConditionNesting_{la+1} \wedge (cd_2, cd) = (cd''_2, cd''_1)$$
$$\Rightarrow cd_1 = cd_2)$$
$$\wedge ((cd_2, cd) \in ConditionNesting_{la+1} \wedge (cd_1, cd) = (cd''_2, cd''_1)$$
$$\Rightarrow cd_1 = cd_2)$$

$$\Longleftrightarrow \quad ((cd_1, cd) \in ConditionNesting_{la+1} \wedge (cd_2, cd) \in ConditionNesting_{la+1}$$
$$\Rightarrow cd_1 = cd_2)$$
$$\wedge ((cd_1, cd) \in ConditionNesting_{la+1} \wedge (cd_2, cd) = (cd''_2, cd''_1)$$
$$\Rightarrow cd_1 = cd_2)$$
$$\wedge ((cd_1, cd) = (cd''_2, cd''_1) \wedge (cd_2, cd) \in ConditionNesting_{la+1})$$
$$\Rightarrow cd_1 = cd_2)$$
$$\wedge ((cd_1, cd) = (cd''_2, cd''_1) \wedge (cd_2, cd) = (cd''_2, cd''_1)$$
$$\Rightarrow cd_1 = cd_2)$$

$$\Longleftrightarrow \quad ((cd_1, cd) \in ConditionNesting_{la+1} \vee (cd_1, cd) = (cd''_2, cd''_1))$$
$$\wedge ((cd_2, cd) \in ConditionNesting_{la+1} \vee (cd_2, cd) = (cd''_2, cd''_1))$$
$$\Rightarrow cd_1 = cd_2$$

$$\Longleftrightarrow \quad (cd_1, cd) \in ConditionNesting_{la+1} \cup (cd''_2, cd''_1)$$
$$\wedge (cd_2, cd) \in ConditionNesting_{la+1} \cup (cd''_2, cd''_1)$$
$$\Rightarrow cd_1 = cd_2$$

$$\overset{(4.146)}{\Longleftrightarrow} \quad (cd_1, cd) \in ConditionNesting'_{la+1}$$
$$\wedge (cd_2, cd) \in ConditionNesting_{la+1'}$$
$$\Rightarrow cd_1 = cd_2$$

$$\overset{(3.97)}{\Longleftrightarrow} \quad (3.97) \text{ holds for } PM'_{la+1}$$

(3.98) holds for PM'_{la}

$\overset{(3.98)}{\Longleftrightarrow}$ $(cd_2, cd_1) \in ConditionNesting'_{la}$
$$\Rightarrow cd_1 \notin outerConditionsOfCondition(cd_1)$$

$\overset{(4.142)}{\Longleftrightarrow}$ $(cd_2, cd_1) \in ConditionNesting_{la} \cup (cd'_2, cd'_1)$
$$\Rightarrow cd_1 \notin outerConditionsOfCondition(cd_1)$$

$\Longrightarrow$ $(cd_2, cd_1) \in ConditionNesting_{la}$
$$\Rightarrow cd_1 \notin outerConditionsOfCondition(cd_1)$$

$\overset{I.H.}{\Longrightarrow}$ $(cd_2, cd_1) \in ConditionNesting_{la+1}$
$$\Rightarrow cd_1 \notin outerConditionsOfCondition(cd_1)$$

$\overset{\substack{(3.102),(4.144),(4.148),\\ I.H.,(4.85),(4.103)}}{\Longleftrightarrow}$ $((cd_2, cd_1) \in ConditionNesting_{la+1}$
$$\Rightarrow cd_1 \notin outerConditionsOfCondition(cd_1))$$
$$\wedge ((cd_2, cd_1) = (cd''_2, cd''_1)$$
$$\Rightarrow cd_1 \notin outerConditionsOfCondition(cd_1))$$

$\Longleftrightarrow$ $(cd_2, cd_1) \in ConditionNesting_{la+1} \cup (cd''_2, cd''_1)$
$$\Rightarrow cd_1 \notin outerConditionsOfCondition(cd_1)$$

$\overset{(4.146)}{\Longleftrightarrow}$ $(cd_2, cd_1) \in ConditionNesting'_{la+1}$
$$\Rightarrow cd_1 \notin outerConditionsOfCondition(cd_1)$$

$\overset{(3.98)}{\Longleftrightarrow}$ (3.98) holds for PM'_{la+1}

(3.99) holds for PM'_{la}

$\overset{(3.99)}{\Longleftrightarrow}$ $(cd_2, cd_1) \in ConditionNesting'_{la}$
$\Rightarrow \exists ty.(undef, cd_1, ty) \in Condition'_{la}$

$\overset{(4.142),(4.143)}{\Longleftrightarrow}$ $(cd_2, cd_1) \in ConditionNesting_{la} \cup (cd'_2, cd'_1)$
$\Rightarrow \exists ty.(undef, cd_1, ty) \in Condition_{la}$

$\overset{}{\Longrightarrow}$ $(cd_2, cd_1) \in ConditionNesting_{la}$
$\Rightarrow \exists ty.(undef, cd_1, ty) \in Condition_{la}$

$\overset{\text{I.H.}}{\Longrightarrow}$ $(cd_2, cd_1) \in ConditionNesting_{la+1}$
$\Rightarrow \exists ty.(undef, cd_1, ty) \in Condition_{la+1}$

$\overset{}{\Longrightarrow}$ $(cd_2, cd_1) \in ConditionNesting_{la+1}$
$\Rightarrow \exists ty.(undef, cd_1, ty) \in Condition_{la+1} \cup (undef, cd''_1, ty')$

$\overset{}{\Longleftarrow}$ $((cd_2, cd_1) \in ConditionNesting_{la+1}$
$\Rightarrow \exists ty.(undef, cd_1, ty) \in Condition_{la+1} \cup (undef, cd''_1, ty'))$
$\wedge ((cd_2, cd_1) = (cd''_2, cd''_1)$
$\Rightarrow \exists ty.(undef, cd_1, ty) \in Condition_{la+1} \cup (undef, cd''_1, ty'))$

$\overset{}{\Longleftrightarrow}$ $(cd_2, cd_1) \in ConditionNesting_{la+1} \cup (cd''_2, cd''_1)$
$\Rightarrow \exists ty.(undef, cd_1, ty) \in Condition_{la+1} \cup (undef, cd''_1, ty')$

$\overset{(4.146),(4.147)}{\Longleftrightarrow}$ $(cd_2, cd_1) \in ConditionNesting'_{la+1}$
$\Rightarrow \exists ty.(undef, cd_1, ty) \in Condition'_{la+1}$

$\overset{(3.99)}{\Longleftrightarrow}$ (3.99) holds for PM'_{la+1}

(3.100) holds for PM'_{la}

$\overset{(3.100)}{\Longleftrightarrow}$ $(po, cd, ty) \in Condition'_{la} \wedge ty \in \{eq, gt, ge, lt, le, op\}$
$\Leftrightarrow |innerConditionsOfCondition(cd)| = 0$

$\overset{(4.143)}{\Longleftrightarrow}$ $(po, cd, ty) \in Condition_{la} \wedge ty \in \{eq, gt, ge, lt, le, op\}$
$\Leftrightarrow |innerConditionsOfCondition(cd)| = 0$

$\overset{I.H.}{\Longrightarrow}$ $(po, cd, ty) \in Condition_{la+1} \wedge ty \in \{eq, gt, ge, lt, le, op\}$
$\Leftrightarrow |innerConditionsOfCondition(cd)| = 0$

$\overset{\substack{(3.103),(4.146),\\ I.H.,(4.85)}}{\Longleftrightarrow}$ $((po, cd, ty) \in Condition_{la+1} \wedge ty \in \{eq, gt, ge, lt, le, op\}$
$\Leftrightarrow |innerConditionsOfCondition(cd)| = 0)$
$\wedge ((po, cd, ty) = (undef, cd''_1, ty') \wedge ty \in \{eq, gt, ge, lt, le, op\}$
$\Leftrightarrow |innerConditionsOfCondition(cd)| = 0)$

$\Longleftrightarrow$ $(po, cd, ty) \in Condition_{la+1} \cup (undef, cd''_1, ty') \wedge ty \in \{eq, gt, ge, lt, le, op\}$
$\Leftrightarrow |innerConditionsOfCondition(cd)| = 0$

$\overset{(4.147)}{\Longleftrightarrow}$ $(po, cd, ty) \in Condition'_{la+1} \wedge ty \in \{eq, gt, ge, lt, le, op\}$
$\Leftrightarrow |innerConditionsOfCondition(cd)| = 0$

$\overset{(3.100)}{\Longleftrightarrow}$ (3.100) holds for PM'_{la+1}

(3.101) holds for PM'_{la}

$\overset{(3.101)}{\Longleftrightarrow}$ $(po, cd, ty) \in Condition'_{la} \wedge ty \in \{not\}$
$\Leftrightarrow |innerConditionsOfCondition(cd)| = 1$

$\overset{(4.143)}{\Longleftrightarrow}$ $(po, cd, ty) \in Condition_{la} \wedge ty \in \{not\}$
$\Leftrightarrow |innerConditionsOfCondition(cd)| = 1$

$\overset{\text{I.H.}}{\Longrightarrow}$ $(po, cd, ty) \in Condition_{la+1} \wedge ty \in \{not\}$
$\Leftrightarrow |innerConditionsOfCondition(cd)| = 1$

$\overset{\substack{(3.103),(4.146),\\ \text{I.H.},(4.85)}}{\Longleftrightarrow}$ $((po, cd, ty) \in Condition_{la+1} \wedge ty \in \{not\}$
$\Leftrightarrow |innerConditionsOfCondition(cd)| = 1)$
$\wedge ((po, cd, ty) = (undef, cd''_1, ty') \wedge ty \in \{not\}$
$\Leftrightarrow |innerConditionsOfCondition(cd)| = 1)$

$\Longleftrightarrow$ $(po, cd, ty) \in Condition_{la+1} \cup (undef, cd''_1, ty') \wedge ty \in \{not\}$
$\Leftrightarrow |innerConditionsOfCondition(cd)| = 1$

$\overset{(4.147)}{\Longleftrightarrow}$ $(po, cd, ty) \in Condition'_{la+1} \wedge ty \in \{not\}$
$\Leftrightarrow |innerConditionsOfCondition(cd)| = 1$

$\overset{(3.101)}{\Longleftrightarrow}$ (3.101) holds for PM'_{la+1}

B.5 Extension with an Action

$\qquad$ (3.106) holds for PM'_{la}

$\overset{(3.106)}{\Longleftrightarrow}$ $(po, ac, no) \in Action'_{la} \Rightarrow po \in Po \lor po = undef$

$\overset{(4.149)}{\Longleftrightarrow}$ $(po, ac, no) \in Action_{la} \cup (po', ac', no') \Rightarrow po \in Po \lor po = undef$

$\Longrightarrow$ $(po, ac, no) \in Action_{la} \Rightarrow po \in Po \lor po = undef$

$\overset{\text{I.H.}}{\Longrightarrow}$ $(po, ac, no) \in Action_{la+1} \Rightarrow po \in Po \lor po = undef$

$\overset{(4.150),(4.89),(4.93)}{\Longleftrightarrow}$ $((po, ac, no) \in Action_{la+1} \Rightarrow po \in Po \lor po = undef)$
$\qquad \land po'' \in policiesOfLayer(la + 1)$

$\overset{(3.73),(3.70)}{\Longrightarrow}$ $((po, ac, no) \in Action_{la+1} \Rightarrow po \in Po \lor po = undef)$
$\qquad \land po'' \in Po$

$\Longleftrightarrow$ $((po, ac, no) \in Action_{la+1} \Rightarrow po \in Po \lor po = undef)$
$\qquad \land (po, ac, no) = (po'', ac'', no') \Rightarrow po \in Po \lor po = undef$

$\Longleftrightarrow$ $(po, ac, no) \in Action_{la+1} \cup (po'', ac'', no') \Rightarrow po \in Po \lor po = undef$

$\overset{(4.151)}{\Longleftrightarrow}$ $(po, ac, no) \in Action'_{la+1} \Rightarrow po \in Po \lor po = undef$

$\overset{(3.106)}{\Longleftrightarrow}$ (3.106) holds for PM'_{la+1}

(3.107) holds for PM'_{la}

$\overset{(3.107)}{\Longleftrightarrow}$ $(po, ac_1, no) \in Action'_{la}$
$\wedge (po, ac_2, no) \in Action'_{la}$
$\Rightarrow ac_1 = ac_2$

$\overset{(4.149)}{\Longleftrightarrow}$ $(po, ac_1, no) \in Action_{la} \cup (po', ac', no')$
$\wedge (po, ac_2, no) \in Action_{la} \cup (po', ac', no')$
$\Rightarrow ac_1 = ac_2$

$\overset{}{\Longrightarrow}$ $(po, ac_1, no) \in Action_{la} \wedge (po, ac_2, no) \in Action_{la}$
$\Rightarrow ac_1 = ac_2$

$\overset{I.H.}{\Longrightarrow}$ $(po, ac_1, no) \in Action_{la+1} \wedge (po, ac_2, no) \in Action_{la+1}$
$\Rightarrow ac_1 = ac_2$

$\overset{(4.105),(4.150),(4.73),}{\underset{(4.149),I.H.,(4.151)}{\Longleftrightarrow}}$ $((po, ac_1, no) \in Action_{la+1} \wedge (po, ac_2, no) \in Action_{la+1}$
$\Rightarrow ac_1 = ac_2)$
$\wedge ((po, ac_1, no) \in Action_{la+1} \wedge (po, ac_2, no) = (po'', ac'', no')$
$\Rightarrow ac_1 = ac_2)$
$\wedge ((po, ac_2, no) \in Action_{la+1} \wedge (po, ac_1, no) = (po'', ac'', no')$
$\Rightarrow ac_1 = ac_2)$

$\overset{}{\Longleftrightarrow}$ $((po, ac_1, no) \in Action_{la+1} \wedge (po, ac_2, no) \in Action_{la+1}$
$\Rightarrow ac_1 = ac_2)$
$\wedge ((po, ac_1, no) \in Action_{la+1} \wedge (po, ac_2, no) = (po'', ac'', no')$
$\Rightarrow ac_1 = ac_2)$
$\wedge ((po, ac_1, no) = (po'', ac'', no') \wedge (po, ac_2, no) \in Action_{la+1})$
$\Rightarrow ac_1 = ac_2)$
$\wedge ((po, ac_1, no) = (po'', ac'', no') \wedge (po, ac_2, no) = (po'', ac'', no')$
$\Rightarrow ac_1 = ac_2)$

$\overset{}{\Longleftrightarrow}$ $((po, ac_1, no) \in Action_{la+1} \vee (po, ac_1, no) = (po'', ac'', no'))$
$\wedge ((po, ac_2, no) \in Action_{la+1} \vee (po, ac_2, no) = (po'', ac'', no'))$
$\Rightarrow ac_1 = ac_2$

$\overset{}{\Longleftrightarrow}$ $(po, ac_1, no) \in Action_{la+1} \cup (po'', ac'', no')$
$\wedge (po, ac_2, no) \in Action_{la+1} \cup (po'', ac'', no')$
$\Rightarrow ac_1 = ac_2$

$\overset{(4.151)}{\Longleftrightarrow}$ $(po, ac_1, no) \in Action'_{la+1}$
$\wedge (po, ac_2, no) \in Action'_{la+1}$
$\Rightarrow ac_1 = ac_2$

$\overset{(3.107)}{\Longleftrightarrow}$ (3.107) holds for PM'_{la+1}

B.6 Extension with an Event Link

$$(3.134) \text{ holds for } LM'_{la}$$

$$\overset{(3.134)}{\Longleftrightarrow} \quad (ev, co) \in EL'_{la} \Rightarrow ev \in Ev \wedge co \in Co$$

$$\overset{(4.153)}{\Longleftrightarrow} \quad (ev, co) \in EL_{la} \cup (ev', co') \Rightarrow ev \in Ev \wedge co \in Co$$

$$\Longrightarrow \quad (ev, co) \in EL_{la} \Rightarrow ev \in Ev \wedge co \in Co$$

$$\overset{\text{I.H.}}{\Longrightarrow} \quad (ev, co) \in EL_{la+1} \Rightarrow ev \in Ev \wedge co \in Co$$

$$\overset{(4.154),(4.77),(4.81)}{\Longleftrightarrow} \quad ((ev, co) \in EL_{la+1} \Rightarrow ev \in Ev \wedge co \in Co)$$
$$\wedge\, ev'' \in eventsOfLayer(la + 1)$$

$$\overset{(3.81),(3.77)}{\Longrightarrow} \quad ((ev, co) \in EL_{la+1} \Rightarrow ev \in Ev \wedge co \in Co)$$
$$\wedge\, ev'' \in Ev$$

$$\overset{\substack{(4.155),(4.36),(4.3),\\(4.6),(4.156)}}{\Longleftrightarrow} \quad ((ev, co) \in EL_{la+1} \Rightarrow ev \in Ev \wedge co \in Co)$$
$$\wedge\, ev'' \in Ev$$
$$\wedge\, co''_i \in conceptsOfLayer(la + 1)$$

$$\overset{(3.17),(3.15)}{\Longrightarrow} \quad ((ev, co) \in EL_{la+1} \Rightarrow ev \in Ev \wedge co \in Co)$$
$$\wedge\, ev'' \in Ev$$
$$\wedge\, co''_i \in Co$$

$$\Longleftrightarrow \quad ((ev, co) \in EL_{la+1} \Rightarrow ev \in Ev \wedge co \in Co)$$
$$\wedge\, ((ev, co) = (ev'', co''_i) \Rightarrow ev \in Ev \wedge co \in Co)$$

$$\Longleftrightarrow \quad (ev, co) \in EL_{la+1} \cup (ev'', co''_i) \Rightarrow ev \in Ev \wedge co \in Co$$

$$\overset{(4.156)}{\Longleftrightarrow} \quad (ev, co) \in EL'_{la+1} \Rightarrow ev \in Ev \wedge co \in Co$$

$$\overset{(3.134)}{\Longleftrightarrow} \quad (3.134) \text{ holds for } LM'_{la+1}$$

$$(3.135) \text{ holds for } LM'_{la}$$

$$\overset{(3.135)}{\Longleftrightarrow} \quad (ev, co_1) \in EL'_{la}$$
$$\wedge (ev, co_2) \in EL'_{la}$$
$$\Rightarrow co_1 = co_2$$

$$\overset{(4.153)}{\Longleftrightarrow} \quad (ev, co_1) \in EL_{la} \cup (ev', co')$$
$$\wedge (ev, co_2) \in EL_{la} \cup (ev', co')$$
$$\Rightarrow co_1 = co_2$$

$$\Longrightarrow \quad (ev, co_1) \in EL_{la} \wedge (ev, co_2) \in EL_{la}$$
$$\Rightarrow co_1 = co_2$$

$$\overset{\text{I.H.}}{\Longrightarrow} \quad (ev, co_1) \in EL_{la+1} \wedge (ev, co_2) \in EL_{la+1}$$
$$\Rightarrow co_1 = co_2$$

$$\overset{(4.105),(4.154),(4.79),}{\underset{(4.153),\text{I.H.},(4.156)}{\Longleftrightarrow}} \quad ((ev, co_1) \in EL_{la+1} \wedge (ev, co_2) \in EL_{la+1}$$
$$\Rightarrow co_1 = co_2)$$
$$\wedge ((ev, co_1) \in EL_{la+1} \wedge (ev, co_2) = (ev'', co''_i)$$
$$\Rightarrow co_1 = co_2)$$
$$\wedge ((ev, co_2) \in EL_{la+1} \wedge (ev, co_1) = (ev'', co''_i)$$
$$\Rightarrow co_1 = co_2)$$

$$\Longleftrightarrow \quad ((ev, co_1) \in EL_{la+1} \wedge (ev, co_2) \in EL_{la+1}$$
$$\Rightarrow co_1 = co_2)$$
$$\wedge ((ev, co_1) \in EL_{la+1} \wedge (ev, co_2) = (ev'', co''_i)$$
$$\Rightarrow co_1 = co_2)$$
$$\wedge ((ev, co_1) = (ev'', co''_i) \wedge (ev, co_2) \in EL_{la+1}$$
$$\Rightarrow co_1 = co_2)$$
$$\wedge ((ev, co_1) = (ev'', co''_i) \wedge (ev, co_2) = (ev'', co''_i)$$
$$\Rightarrow co_1 = co_2)$$

$$\Longleftrightarrow \quad ((ev, co_1) \in EL_{la+1} \vee (ev, co_1) = (ev'', co''_i))$$
$$\wedge ((ev, co_2) \in EL_{la+1} \vee (ev, co_2) = (ev'', co''_i))$$
$$\Rightarrow co_1 = co_2$$

$$\Longleftrightarrow \quad (ev, co_1) \in EL_{la+1} \cup (ev'', co''_i)$$
$$\wedge (ev, co_2) \in EL_{la+1} \cup (ev'', co''_i)$$
$$\Rightarrow co_1 = co_2$$

$$\overset{(4.156)}{\Longleftrightarrow} \quad (ev, co_1) \in EL'_{la+1}$$
$$\wedge (ev, co_2) \in EL'_{la+1}$$
$$\Rightarrow co_1 = co_2$$

$$\overset{(3.135)}{\Longleftrightarrow} \quad (3.135) \text{ holds for } LM'_{la+1}$$

(3.136) holds for LM'_{la}

$\overset{(3.136)}{\Longleftrightarrow}$ $(ev, co) \in EL'_{la}$

$\Rightarrow layersOfEvent(ev) \subseteq layersOfConcept(co)$

$\overset{(4.153)}{\Longleftrightarrow}$ $(ev, co) \in EL_{la} \cup (ev', co')$

$\Rightarrow layersOfEvent(ev) \subseteq layersOfConcept(co)$

$\Longrightarrow$ $(ev, co) \in EL_{la}$

$\Rightarrow layersOfEvent(ev) \subseteq layersOfConcept(co)$

$\overset{\text{I.H.}}{\Longrightarrow}$ $(ev, co) \in EL_{la+1}$

$\Rightarrow layersOfEvent(ev) \subseteq layersOfConcept(co)$

$\begin{array}{c} {\scriptstyle (4.155),(4.36),(4.3),} \\ {\scriptstyle (4.6),(4.156),} \\ {\scriptstyle (3.17),(3.16)} \\ \Longleftrightarrow \end{array}$ $((ev, co) \in EL_{la+1}$

$\Rightarrow layersOfEvent(ev) \subseteq layersOfConcept(co))$

$\wedge \, layersOfEvent(ev'') \subseteq layersOfConcept(co''_i)$

$\Longleftrightarrow$ $((ev, co) \in EL_{la+1}$

$\Rightarrow layersOfEvent(ev) \subseteq layersOfConcept(co))$

$\wedge \, ((ev, co) = (ev'', co''_i)$

$\Rightarrow layersOfEvent(ev) \subseteq layersOfConcept(co))$

$\Longleftrightarrow$ $(ev, co) \in EL_{la+1} \cup (ev'', co''_i)$

$\Rightarrow layersOfEvent(ev) \subseteq layersOfConcept(co)$

$\overset{(4.156)}{\Longleftrightarrow}$ $(ev, co) \in EL'_{la+1}$

$\Rightarrow layersOfEvent(ev) \subseteq layersOfConcept(co)$

$\overset{(3.136)}{\Longleftrightarrow}$ (3.136) holds for LM'_{la+1}

B.7 Extension with a Binary Expression Link

We prove the semantical correctness for the case of two literals as arguments of the binary expression link, according to section 4.2.3, case 7a. The other cases 7b to 7i can be proved accordingly.

(3.143) holds for LM'_{la}

$$\overset{(3.143)}{\Longleftrightarrow} (be, arg_1, arg_2) \in BEL'_{la} \Rightarrow be \in Be$$
$$\wedge (arg_1 \in Id_{Pr} \Rightarrow arg_1 \in Pr) \wedge (arg_2 \in Id_{Pr} \Rightarrow arg_2 \in Pr)$$
$$\wedge (arg_1 \in Id_{OI} \Rightarrow arg_1 \in OIL1) \wedge (arg_2 \in Id_{OI} \Rightarrow arg_2 \in OIL1)$$

$$\overset{(4.157)}{\Longleftrightarrow} (be, arg_1, arg_2) \in BEL_{la} \cup (be', arg'_1, arg'_2) \Rightarrow be \in Be$$
$$\wedge (arg_1 \in Id_{Pr} \Rightarrow arg_1 \in Pr) \wedge (arg_2 \in Id_{Pr} \Rightarrow arg_2 \in Pr)$$
$$\wedge (arg_1 \in Id_{OI} \Rightarrow arg_1 \in OIL1) \wedge (arg_2 \in Id_{OI} \Rightarrow arg_2 \in OIL1)$$

$$\Longrightarrow (be, arg_1, arg_2) \in BEL_{la} \Rightarrow be \in Be$$
$$\wedge (arg_1 \in Id_{Pr} \Rightarrow arg_1 \in Pr) \wedge (arg_2 \in Id_{Pr} \Rightarrow arg_2 \in Pr)$$
$$\wedge (arg_1 \in Id_{OI} \Rightarrow arg_1 \in OIL1) \wedge (arg_2 \in Id_{OI} \Rightarrow arg_2 \in OIL1)$$

$$\overset{I.H.}{\Longrightarrow} (be, arg_1, arg_2) \in BEL_{la+1} \Rightarrow be \in Be$$
$$\wedge (arg_1 \in Id_{Pr} \Rightarrow arg_1 \in Pr) \wedge (arg_2 \in Id_{Pr} \Rightarrow arg_2 \in Pr)$$
$$\wedge (arg_1 \in Id_{OI} \Rightarrow arg_1 \in OIL1) \wedge (arg_2 \in Id_{OI} \Rightarrow arg_2 \in OIL1)$$

$$\overset{(4.158),(4.83),(4.87)}{\Longleftrightarrow} ((be, arg_1, arg_2) \in BEL_{la+1} \Rightarrow be \in Be$$
$$\wedge (arg_1 \in Id_{Pr} \Rightarrow arg_1 \in Pr) \wedge (arg_2 \in Id_{Pr} \Rightarrow arg_2 \in Pr)$$
$$\wedge (arg_1 \in Id_{OI} \Rightarrow arg_1 \in OIL1) \wedge (arg_2 \in Id_{OI} \Rightarrow arg_2 \in OIL1))$$
$$\wedge be'' \in conditionsOfLayer(la+1)$$

$$\overset{(4.157),I.H.,(4.102)}{\Longleftrightarrow} ((be, arg_1, arg_2) \in BEL_{la+1} \Rightarrow be \in Be$$
$$\wedge (arg_1 \in Id_{Pr} \Rightarrow arg_1 \in Pr) \wedge (arg_2 \in Id_{Pr} \Rightarrow arg_2 \in Pr)$$
$$\wedge (arg_1 \in Id_{OI} \Rightarrow arg_1 \in OIL1) \wedge (arg_2 \in Id_{OI} \Rightarrow arg_2 \in OIL1))$$
$$\wedge be'' \in conditionsOfLayer(la+1)$$
$$\wedge typeOfCondition(be'') \in \{eq, gt, ge, lt, le\}$$

$$\overset{(3.93),(3.141)}{\Longrightarrow} ((be, arg_1, arg_2) \in BEL_{la+1} \Rightarrow be \in Be$$
$$\wedge (arg_1 \in Id_{Pr} \Rightarrow arg_1 \in Pr) \wedge (arg_2 \in Id_{Pr} \Rightarrow arg_2 \in Pr)$$
$$\wedge (arg_1 \in Id_{OI} \Rightarrow arg_1 \in OIL1) \wedge (arg_2 \in Id_{OI} \Rightarrow arg_2 \in OIL1))$$
$$\wedge be'' \in Be$$

$$\overset{(4.157),I.H.}{\Longrightarrow} ((be, arg_1, arg_2) \in BEL_{la+1} \Rightarrow be \in Be$$
$$\wedge (arg_1 \in Id_{Pr} \Rightarrow arg_1 \in Pr) \wedge (arg_2 \in Id_{Pr} \Rightarrow arg_2 \in Pr)$$
$$\wedge (arg_1 \in Id_{OI} \Rightarrow arg_1 \in OIL1) \wedge (arg_2 \in Id_{OI} \Rightarrow arg_2 \in OIL1))$$
$$\wedge (be'' \in Be$$
$$\wedge (arg'_1 \in Id_{Pr} \Rightarrow arg'_1 \in Pr) \wedge (arg'_2 \in Id_{Pr} \Rightarrow arg'_2 \in Pr)$$
$$\wedge (arg'_1 \in Id_{OI} \Rightarrow arg'_1 \in OIL1) \wedge (arg'_2 \in Id_{OI} \Rightarrow arg'_2 \in OIL1))$$

$\iff$ $\quad((be, arg_1, arg_2) \in BEL_{la+1} \Rightarrow be \in Be$
$\qquad \wedge (arg_1 \in Id_{Pr} \Rightarrow arg_1 \in Pr) \wedge (arg_2 \in Id_{Pr} \Rightarrow arg_2 \in Pr)$
$\qquad \wedge (arg_1 \in Id_{OI} \Rightarrow arg_1 \in OIL1) \wedge (arg_2 \in Id_{OI} \Rightarrow arg_2 \in OIL1))$
$\qquad \wedge ((be, arg_1, arg_2) = (be'', arg_1', arg_2') \Rightarrow be \in Be$
$\qquad \wedge (arg_1 \in Id_{Pr} \Rightarrow arg_1 \in Pr) \wedge (arg_2 \in Id_{Pr} \Rightarrow arg_2 \in Pr)$
$\qquad \wedge (arg_1 \in Id_{OI} \Rightarrow arg_1 \in OIL1) \wedge (arg_2 \in Id_{OI} \Rightarrow arg_2 \in OIL1))$

$\iff$ $\quad(be, arg_1, arg_2) \in BEL_{la+1} \cup (be'', arg_1', arg_2') \Rightarrow be \in Be$
$\qquad \wedge (arg_1 \in Id_{Pr} \Rightarrow arg_1 \in Pr) \wedge (arg_2 \in Id_{Pr} \Rightarrow arg_2 \in Pr)$
$\qquad \wedge (arg_1 \in Id_{OI} \Rightarrow arg_1 \in OIL1) \wedge (arg_2 \in Id_{OI} \Rightarrow arg_2 \in OIL1)$

$\overset{(4.160)}{\iff}$ $\quad(be, arg_1, arg_2) \in BEL'_{la+1} \Rightarrow be \in Be$
$\qquad \wedge (arg_1 \in Id_{Pr} \Rightarrow arg_1 \in Pr) \wedge (arg_2 \in Id_{Pr} \Rightarrow arg_2 \in Pr)$
$\qquad \wedge (arg_1 \in Id_{OI} \Rightarrow arg_1 \in OIL1) \wedge (arg_2 \in Id_{OI} \Rightarrow arg_2 \in OIL1)$

$\overset{(3.143)}{\iff}$ $\quad(3.143)$ holds for LM'_{la+1}

$$(3.144) \text{ holds for } LM'_{la}$$

$$\overset{(3.144)}{\Longleftrightarrow} \quad (be, arg_1, arg_2) \in BEL'_{la}$$
$$\wedge (be, arg_3, arg_4) \in BEL'_{la}$$
$$\Rightarrow arg_1 = arg_3 \wedge arg_2 = arg_4$$

$$\overset{(4.157)}{\Longleftrightarrow} \quad (be, arg_1, arg_2) \in BEL_{la} \cup (be', arg'_1, arg'_2)$$
$$\wedge (be, arg_3, arg_4) \in BEL_{la} \cup (be', arg'_1, arg'_2)$$
$$\Rightarrow arg_1 = arg_3 \wedge arg_2 = arg_4$$

$$\Longrightarrow \quad (be, arg_1, arg_2) \in BEL_{la}$$
$$\wedge (be, arg_3, arg_4) \in BEL_{la}$$
$$\Rightarrow arg_1 = arg_3 \wedge arg_2 = arg_4$$

$$\overset{\text{I.H.}}{\Longrightarrow} \quad (be, arg_1, arg_2) \in BEL_{la+1}$$
$$\wedge (be, arg_3, arg_4) \in BEL_{la+1}$$
$$\Rightarrow arg_1 = arg_3 \wedge arg_2 = arg_4$$

$$\overset{\substack{(4.106),(4.158),(4.85),\\(4.157),\text{I.H.}}}{\Longleftrightarrow} \quad ((be, arg_1, arg_2) \in BEL_{la+1} \wedge (be, arg_3, arg_4) \in BEL_{la+1}$$
$$\Rightarrow arg_1 = arg_3 \wedge arg_2 = arg_4)$$
$$\wedge ((be, arg_1, arg_2) \in BEL_{la+1} \wedge (be, arg_3, arg_4) = (be'', arg'_1, arg'_2)$$
$$\Rightarrow arg_1 = arg_3 \wedge arg_2 = arg_4)$$
$$\wedge ((be, arg_3, arg_4) \in BEL_{la+1} \wedge (be, arg_1, arg_2) = (be'', arg'_1, arg'_2)$$
$$\Rightarrow arg_1 = arg_3 \wedge arg_2 = arg_4)$$

$$\Longleftrightarrow \quad ((be, arg_1, arg_2) \in BEL_{la+1} \wedge (be, arg_3, arg_4) \in BEL_{la+1}$$
$$\Rightarrow arg_1 = arg_3 \wedge arg_2 = arg_4)$$
$$\wedge ((be, arg_1, arg_2) \in BEL_{la+1} \wedge (be, arg_3, arg_4) = (be'', arg'_1, arg'_2)$$
$$\Rightarrow arg_1 = arg_3 \wedge arg_2 = arg_4)$$
$$\wedge ((be, arg_1, arg_2) = (be'', arg'_1, arg'_2) \wedge (be, arg_3, arg_4) \in BEL_{la+1}$$
$$\Rightarrow arg_1 = arg_3 \wedge arg_2 = arg_4)$$
$$\wedge ((be, arg_1, arg_2) = (be'', arg'_1, arg'_2) \wedge (be, arg_3, arg_4) = (be'', arg'_1, arg'_2)$$
$$\Rightarrow arg_1 = arg_3 \wedge arg_2 = arg_4)$$

$$\Longleftrightarrow \quad ((be, arg_1, arg_2) \in BEL_{la+1} \vee (be, arg_1, arg_2) = (be'', arg'_1, arg'_2))$$
$$\wedge ((be, arg_3, arg_4) \in BEL_{la+1} \vee (be, arg_3, arg_4) = (be'', arg'_1, arg'_2))$$
$$\Rightarrow arg_1 = arg_3 \wedge arg_2 = arg_4$$

$$\Longleftrightarrow \quad (be, arg_1, arg_2) \in BEL_{la+1} \cup (be'', arg'_1, arg'_2)$$
$$\wedge (be, arg_3, arg_4) \in BEL_{la+1} \cup (be'', arg'_1, arg'_2)$$
$$\Rightarrow arg_1 = arg_3 \wedge arg_2 = arg_4$$

$$\overset{(4.160)}{\Longleftrightarrow} \quad (be, arg_1, arg_2) \in BEL'_{la+1}$$
$$\wedge (be, arg_3, arg_4) \in BEL'_{la+1}$$
$$\Rightarrow arg_1 = arg_3 \wedge arg_2 = arg_4$$

$$\overset{(3.144)}{\Longleftrightarrow} \quad (3.144) \text{ holds for } LM'_{la+1}$$

(3.145) holds for LM'_{la}

$\overset{(3.145)}{\Longleftrightarrow}$ $(be, arg_1, arg_2) \in BEL'_{la}$

$\Rightarrow (arg_1 \in Id_{Pr} \Rightarrow \forall po \in policiesOfCondition(be).$
$arg_1 \in visibleProperties(po))$
$\wedge (arg_2 \in Id_{Pr} \Rightarrow \forall po \in policiesOfCondition(be).$
$arg_2 \in visibleProperties(po))$

$\overset{(4.157)}{\Longleftrightarrow}$ $(be, arg_1, arg_2) \in BEL_{la} \cup (be', arg'_1, arg'_2)$

$\Rightarrow (arg_1 \in Id_{Pr} \Rightarrow \forall po \in policiesOfCondition(be).$
$arg_1 \in visibleProperties(po))$
$\wedge (arg_2 \in Id_{Pr} \Rightarrow \forall po \in policiesOfCondition(be).$
$arg_2 \in visibleProperties(po))$

$\Longrightarrow$ $(be, arg_1, arg_2) \in BEL_{la}$

$\Rightarrow (arg_1 \in Id_{Pr} \Rightarrow \forall po \in policiesOfCondition(be).$
$arg_1 \in visibleProperties(po))$
$\wedge (arg_2 \in Id_{Pr} \Rightarrow \forall po \in policiesOfCondition(be).$
$arg_2 \in visibleProperties(po))$

$\overset{I.H.}{\Longrightarrow}$ $(be, arg_1, arg_2) \in BEL_{la+1}$

$\Rightarrow (arg_1 \in Id_{Pr} \Rightarrow \forall po \in policiesOfCondition(be).$
$arg_1 \in visibleProperties(po))$
$\wedge (arg_2 \in Id_{Pr} \Rightarrow \forall po \in policiesOfCondition(be).$
$arg_2 \in visibleProperties(po))$

$\overset{(4.159),(3.9)}{\Longleftrightarrow}$ $((be, arg_1, arg_2) \in BEL_{la+1}$

$\Rightarrow (arg_1 \in Id_{Pr} \Rightarrow \forall po \in policiesOfCondition(be).$
$arg_1 \in visibleProperties(po))$
$\wedge (arg_2 \in Id_{Pr} \Rightarrow \forall po \in policiesOfCondition(be).$
$arg_2 \in visibleProperties(po)))$
$\wedge ((arg'_1 \in Id_{Pr} \Rightarrow \forall po \in policiesOfCondition(be'').$
$arg_1 \in visibleProperties(po))$
$\wedge (arg'_2 \in Id_{Pr} \Rightarrow \forall po \in policiesOfCondition(be'').$
$arg_2 \in visibleProperties(po)))$

$\Longleftrightarrow$ $((be, arg_1, arg_2) \in BEL_{la+1}$

$\Rightarrow (arg_1 \in Id_{Pr} \Rightarrow \forall po \in policiesOfCondition(be).$
$arg_1 \in visibleProperties(po))$
$\wedge (arg_2 \in Id_{Pr} \Rightarrow \forall po \in policiesOfCondition(be).$
$arg_2 \in visibleProperties(po)))$
$\wedge ((be, arg_1, arg_2) = (be'', arg'_1, arg'_2)$
$\Rightarrow (arg_1 \in Id_{Pr} \Rightarrow \forall po \in policiesOfCondition(be).$
$arg_1 \in visibleProperties(po))$
$\wedge (arg_2 \in Id_{Pr} \Rightarrow \forall po \in policiesOfCondition(be).$
$arg_2 \in visibleProperties(po)))$

$\iff \quad (be, arg_1, arg_2) \in BEL_{la+1} \cup (be'', arg_1', arg_2')$

$\quad \Rightarrow (arg_1 \in Id_{Pr} \Rightarrow \forall po \in policiesOfCondition(be).$

$\qquad arg_1 \in visibleProperties(po))$

$\quad \wedge (arg_2 \in Id_{Pr} \Rightarrow \forall po \in policiesOfCondition(be).$

$\qquad arg_2 \in visibleProperties(po))$

$\overset{(4.160)}{\iff} \quad (be, arg_1, arg_2) \in BEL'_{la+1}$

$\quad \Rightarrow (arg_1 \in Id_{Pr} \Rightarrow \forall po \in policiesOfCondition(be).$

$\qquad arg_1 \in visibleProperties(po))$

$\quad \wedge (arg_2 \in Id_{Pr} \Rightarrow \forall po \in policiesOfCondition(be).$

$\qquad arg_2 \in visibleProperties(po))$

$\overset{(3.145)}{\iff} \quad$ (3.145) holds for LM'_{la+1}

(3.146) holds for LM'_{la}

$\overset{(3.146)}{\Longleftrightarrow}$ $(be, arg_1, arg_2) \in BEL'_{la}$

$\Rightarrow (arg_1 \in Id_{OI} \Rightarrow \forall po \in policiesOfCondition(be).$
$\forall pr \in propertiesOfOpInv(arg_1).pr \in visibleProperties(po))$
$\wedge (arg_2 \in Id_{OI} \Rightarrow \forall po \in policiesOfCondition(be).$
$\forall pr \in propertiesOfOpInv(arg_2).pr \in visibleProperties(po))$

$\overset{(4.157)}{\Longleftrightarrow}$ $(be, arg_1, arg_2) \in BEL_{la} \cup (be', arg'_1, arg'_2)$

$\Rightarrow (arg_1 \in Id_{OI} \Rightarrow \forall po \in policiesOfCondition(be).$
$\forall pr \in propertiesOfOpInv(arg_1).pr \in visibleProperties(po))$
$\wedge (arg_2 \in Id_{OI} \Rightarrow \forall po \in policiesOfCondition(be).$
$\forall pr \in propertiesOfOpInv(arg_2).pr \in visibleProperties(po))$

$\Longrightarrow$ $(be, arg_1, arg_2) \in BEL_{la}$

$\Rightarrow (arg_1 \in Id_{OI} \Rightarrow \forall po \in policiesOfCondition(be).$
$\forall pr \in propertiesOfOpInv(arg_1).pr \in visibleProperties(po))$
$\wedge (arg_2 \in Id_{OI} \Rightarrow \forall po \in policiesOfCondition(be).$
$\forall pr \in propertiesOfOpInv(arg_2).pr \in visibleProperties(po))$

$\overset{I.H.}{\Longrightarrow}$ $(be, arg_1, arg_2) \in BEL_{la+1}$

$\Rightarrow (arg_1 \in Id_{OI} \Rightarrow \forall po \in policiesOfCondition(be).$
$\forall pr \in propertiesOfOpInv(arg_1).pr \in visibleProperties(po))$
$\wedge (arg_2 \in Id_{OI} \Rightarrow \forall po \in policiesOfCondition(be).$
$\forall pr \in propertiesOfOpInv(arg_2).pr \in visibleProperties(po))$

$\overset{(4.159),(3.9)}{\Longleftrightarrow}$ $((be, arg_1, arg_2) \in BEL_{la+1}$

$\Rightarrow (arg_1 \in Id_{OI} \Rightarrow \forall po \in policiesOfCondition(be).$
$\forall pr \in propertiesOfOpInv(arg_1).pr \in visibleProperties(po))$
$\wedge (arg_2 \in Id_{OI} \Rightarrow \forall po \in policiesOfCondition(be).$
$\forall pr \in propertiesOfOpInv(arg_2).pr \in visibleProperties(po)))$
$\wedge ((arg'_1 \in Id_{OI} \Rightarrow \forall po \in policiesOfCondition(be'').$
$\forall pr \in propertiesOfOpInv(arg_1).pr \in visibleProperties(po))$
$\wedge (arg'_2 \in Id_{OI} \Rightarrow \forall po \in policiesOfCondition(be'').$
$\forall pr \in propertiesOfOpInv(arg_2).pr \in visibleProperties(po)))$

$\Longleftrightarrow$ $((be, arg_1, arg_2) \in BEL_{la+1}$

$\Rightarrow (arg_1 \in Id_{OI} \Rightarrow \forall po \in policiesOfCondition(be).$
$\forall pr \in propertiesOfOpInv(arg_1).pr \in visibleProperties(po))$
$\wedge (arg_2 \in Id_{OI} \Rightarrow \forall po \in policiesOfCondition(be).$
$\forall pr \in propertiesOfOpInv(arg_2).pr \in visibleProperties(po)))$
$\wedge ((be, arg_1, arg_2) = (be'', arg'_1, arg'_2)$
$\Rightarrow (arg_1 \in Id_{OI} \Rightarrow \forall po \in policiesOfCondition(be).$
$\forall pr \in propertiesOfOpInv(arg_1).pr \in visibleProperties(po))$
$\wedge (arg_2 \in Id_{OI} \Rightarrow \forall po \in policiesOfCondition(be).$
$\forall pr \in propertiesOfOpInv(arg_2).pr \in visibleProperties(po)))$

$\Longleftarrow \quad (be, arg_1, arg_2) \in BEL_{la+1} \cup (be'', arg_1', arg_2')$

$\quad\quad \Rightarrow (arg_1 \in Id_{OI} \Rightarrow \forall po \in policiesOfCondition(be).$

$\quad\quad\quad\quad \forall pr \in propertiesOfOpInv(arg_1).pr \in visibleProperties(po))$

$\quad\quad\quad \wedge (arg_2 \in Id_{OI} \Rightarrow \forall po \in policiesOfCondition(be).$

$\quad\quad\quad\quad \forall pr \in propertiesOfOpInv(arg_2).pr \in visibleProperties(po))$

$\overset{(4.160)}{\Longleftrightarrow} \quad (be, arg_1, arg_2) \in BEL_{la+1}'$

$\quad\quad \Rightarrow (arg_1 \in Id_{OI} \Rightarrow \forall po \in policiesOfCondition(be).$

$\quad\quad\quad\quad \forall pr \in propertiesOfOpInv(arg_1).pr \in visibleProperties(po))$

$\quad\quad\quad \wedge (arg_2 \in Id_{OI} \Rightarrow \forall po \in policiesOfCondition(be).$

$\quad\quad\quad\quad \forall pr \in propertiesOfOpInv(arg_2).pr \in visibleProperties(po))$

$\overset{(3.146)}{\Longleftrightarrow} \quad (3.146) \text{ holds for } LM_{la+1}'$

(3.147) holds for LM'_{la}

$\underset{(3.147)}{\Longleftrightarrow}$ $(be, arg_1, arg_2) \in BEL'_{la} \Rightarrow layersOfCondition(be)$

$$\subseteq \bigcap_{pr \in propertiesOfBinExpr(be)} layersOfProperty(pr)$$

$$\cap \bigcap_{oi \in opInvsOfBinExpr(be)} layersOfOpInv(oi)$$

$\underset{(4.157)}{\Longleftrightarrow}$ $(be, arg_1, arg_2) \in BEL_{la} \cup (be', arg'_1, arg'_2) \Rightarrow layersOfCondition(be)$

$$\subseteq \bigcap_{pr \in propertiesOfBinExpr(be)} layersOfProperty(pr)$$

$$\cap \bigcap_{oi \in opInvsOfBinExpr(be)} layersOfOpInv(oi)$$

$\Longrightarrow$ $(be, arg_1, arg_2) \in BEL_{la} \Rightarrow layersOfCondition(be)$

$$\subseteq \bigcap_{pr \in propertiesOfBinExpr(be)} layersOfProperty(pr)$$

$$\cap \bigcap_{oi \in opInvsOfBinExpr(be)} layersOfOpInv(oi)$$

$\underset{I.H.}{\Longrightarrow}$ $(be, arg_1, arg_2) \in BEL_{la+1} \Rightarrow layersOfCondition(be)$

$$\subseteq \bigcap_{pr \in propertiesOfBinExpr(be)} layersOfProperty(pr)$$

$$\cap \bigcap_{oi \in opInvsOfBinExpr(be)} layersOfOpInv(oi)$$

$\underset{(4.159),(3.151),(3.152)}{\Longleftrightarrow}$ $((be, arg_1, arg_2) \in BEL_{la+1} \Rightarrow layersOfCondition(be)$

$$\subseteq \bigcap_{pr \in propertiesOfBinExpr(be)} layersOfProperty(pr)$$

$$\cap \bigcap_{oi \in opInvsOfBinExpr(be)} layersOfOpInv(oi))$$

$\wedge layersOfCondition(be'')$

$$\subseteq \bigcap_{pr \in propertiesOfBinExpr(be'')} layersOfProperty(pr)$$

$$\cap \bigcap_{oi \in opInvsOfBinExpr(be'')} layersOfOpInv(oi)$$

$\Longleftrightarrow$ $((be, arg_1, arg_2) \in BEL_{la+1} \Rightarrow layersOfCondition(be)$

$$\subseteq \bigcap_{pr \in propertiesOfBinExpr(be)} layersOfProperty(pr)$$

$$\cap \bigcap_{oi \in opInvsOfBinExpr(be)} layersOfOpInv(oi))$$

$\wedge ((be, arg_1, arg_2) = (be'', arg''_{1_i}, arg''_{2_j})$

$\Rightarrow layersOfCondition(be)$

$$\subseteq \bigcap_{pr \in propertiesOfBinExpr(be)} layersOfProperty(pr)$$

$$\cap \bigcap_{oi \in opInvsOfBinExpr(be)} layersOfOpInv(oi))$$

$$\Longleftrightarrow \quad (be, arg_1, arg_2) \in BEL_{la+1} \cup (be'', arg''_{1_i}, arg''_{2_j}) \Rightarrow layersOfCondition(be)$$

$$\subseteq \bigcap_{pr \in propertiesOfBinExpr(be)} layersOfProperty(pr)$$

$$\cap \bigcap_{oi \in opInvsOfBinExpr(be)} layersOfOpInv(oi)$$

$$\overset{(4.160)}{\Longleftrightarrow} \quad (be, arg_1, arg_2) \in BEL'_{la+1} \Rightarrow layersOfCondition(be)$$

$$\subseteq \bigcap_{pr \in propertiesOfBinExpr(be)} layersOfProperty(pr)$$

$$\cap \bigcap_{oi \in opInvsOfBinExpr(be)} layersOfOpInv(oi)$$

$$\overset{(3.147)}{\Longleftrightarrow} \quad (3.147) \text{ holds for } LM'_{la+1}$$

B.8 Extension with an Operational Expression Link

(3.156) holds for LM'_{la}

$\overset{(3.156)}{\Longleftrightarrow}$ $(oe, oi) \in OEL'_{la} \Rightarrow oe \in Oe \wedge oi \in OIL1$

$\overset{(4.201)}{\Longleftrightarrow}$ $(oe, oi) \in OEL_{la} \cup (oe', oi') \Rightarrow oe \in Oe \wedge oi \in OIL1$

$\Longrightarrow$ $(oe, oi) \in OEL_{la} \Rightarrow oe \in Oe \wedge oi \in OIL1$

$\overset{\text{I.H.}}{\Longrightarrow}$ $(oe, oi) \in OEL_{la+1} \Rightarrow oe \in Oe \wedge oi \in OIL1$

$\overset{(4.202),(4.83),(4.87)}{\Longleftrightarrow}$ $((oe, oi) \in OEL_{la+1} \Rightarrow oe \in Oe \wedge oi \in OIL1)$
$\wedge\, oe'' \in conditionsOfLayer(la + 1)$

$\overset{(4.201),\text{I.H.},(4.102)}{\Longleftrightarrow}$ $((oe, oi) \in OEL_{la+1} \Rightarrow oe \in Oe \wedge oi \in OIL1)$
$\wedge\, oe'' \in conditionsOfLayer(la + 1)$
$\wedge\, typeOfCondition(be'') \in \{op\}$

$\overset{(3.93),(3.154)}{\Longrightarrow}$ $((oe, oi) \in OEL_{la+1} \Rightarrow oe \in Oe \wedge oi \in OIL1)$
$\wedge\, oe'' \in Oe$

$\overset{(4.204),(4.95),(4.99)}{\Longleftrightarrow}$ $((oe, oi) \in OEL_{la+1} \Rightarrow oe \in Oe \wedge oi \in OIL1)$
$\wedge\, oe'' \in Oe$
$\wedge\, oi'' \in opInvsOfLayer(la + 1)$

$\overset{(3.129)}{\Longrightarrow}$ $((oe, oi) \in OEL_{la+1} \Rightarrow oe \in Oe \wedge oi \in OIL1)$
$\wedge\, oe'' \in Oe$
$\wedge\, oi'' \in OIL1$

$\Longleftrightarrow$ $((oe, oi) \in OEL_{la+1} \Rightarrow oe \in Oe \wedge oi \in OIL1)$
$\wedge\, ((oe, oi) = (oe'', oi'') \Rightarrow oe \in Oe \wedge oi \in OIL1)$

$\Longleftrightarrow$ $(oe, oi) \in OEL_{la+1} \cup (oe'', oi'') \Rightarrow oe \in Oe \wedge oi \in OIL1$

$\overset{(4.203)}{\Longleftrightarrow}$ $(oe, oi) \in OEL'_{la+1} \Rightarrow oe \in Oe \wedge oi \in OIL1$

$\overset{(3.156)}{\Longleftrightarrow}$ (3.156) holds for LM'_{la+1}

(3.157) holds for LM'_{la}

$\overset{(3.157)}{\Longleftrightarrow}$ $(oe, oi_1) \in OEL'_{la}$
$\wedge (oe, oi_2) \in OEL'_{la}$
$\Rightarrow oi_1 = oi_2$

$\overset{(4.201)}{\Longleftrightarrow}$ $(oe, oi_1) \in OEL_{la} \cup (oe', oi')$
$\wedge (oe, oi_2) \in OEL_{la} \cup (oe', oi')$
$\Rightarrow oi_1 = oi_2$

$\Longrightarrow$ $(oe, oi_1) \in OEL_{la} \wedge (oe, oi_2) \in OEL_{la}$
$\Rightarrow oi_1 = oi_2$

$\overset{\text{I.H.}}{\Longrightarrow}$ $(oe, oi_1) \in OEL_{la+1} (oe, oi_2) \in OEL_{la+1}$
$\Rightarrow oi_1 = oi_2$

$(4.115),(4.202),(4.85),$
$(4.201),\text{I.H.},(4.203)$
$\Longleftrightarrow$
$((oe, oi_1) \in OEL_{la+1} \wedge (oe, oi_2) \in OEL_{la+1}$
$\Rightarrow oi_1 = oi_2)$
$\wedge ((oe, oi_1) \in OEL_{la+1} \wedge (oe, oi_2) = (oe'', oi'')$
$\Rightarrow oi_1 = oi_2)$
$\wedge ((oe, oi_2) \in OEL_{la+1} \wedge (oe, oi_1) = (oe'', oi'')$
$\Rightarrow oi_1 = oi_2)$

$\Longleftrightarrow$ $((oe, oi_1) \in OEL_{la+1} \wedge (oe, oi_2) \in OEL_{la+1}$
$\Rightarrow oi_1 = oi_2)$
$\wedge ((oe, oi_1) \in OEL_{la+1} \wedge (oe, oi_2) = (oe'', oi'')$
$\Rightarrow oi_1 = oi_2)$
$\wedge ((oe, oi_1) = (oe'', oi'') \wedge (oe, oi_2) \in OEL_{la+1}$
$\Rightarrow oi_1 = oi_2)$
$\wedge ((oe, oi_1) = (oe'', oi'') \wedge (oe, oi_2) = (oe'', oi'')$
$\Rightarrow co_1 = co_2)$

$\Longleftrightarrow$ $((oe, oi_1) \in OEL_{la+1} \vee (oe, oi_1) = (oe'', oi''))$
$\wedge ((oe, oi_2) \in OEL_{la+1} \vee (oe, oi_2) = (oe'', oi''))$
$\Rightarrow oi_1 = oi_2$

$\Longleftrightarrow$ $(oe, oi_1) \in OEL_{la+1} \cup (oe'', oi'')$
$\wedge (oe, oi_2) \in OEL_{la+1} \cup (oe'', oi'')$
$\Rightarrow oi_1 = oi_2$

$\overset{(4.203)}{\Longleftrightarrow}$ $(oe, oi_1) \in OEL'_{la+1}$
$\wedge (oe, oi_2) \in OEL'_{la+1}$
$\Rightarrow oi_1 = oi_2$

$\overset{(3.157)}{\Longleftrightarrow}$ (3.157) holds for LM'_{la+1}

(3.158) holds for LM'_{la}

$\overset{(3.158)}{\Longleftrightarrow}$ $(oe, oi) \in OEL'_{la} \Rightarrow \forall po \in policiesOfCondition(oe)$
$\forall pr \in propertiesOfOpInv(oi).pr \in visibleProperties(po)$

$\overset{(4.201)}{\Longleftrightarrow}$ $(oe, oi) \in OEL_{la} \cup (oe', oi') \Rightarrow \forall po \in policiesOfCondition(oe)$
$\forall pr \in propertiesOfOpInv(oi).pr \in visibleProperties(po)$

$\Longrightarrow$ $(oe, oi) \in OEL_{la} \Rightarrow \forall po \in policiesOfCondition(oe)$
$\forall pr \in propertiesOfOpInv(oi).pr \in visibleProperties(po)$

$\overset{\text{I.H.}}{\Longrightarrow}$ $(oe, oi) \in OEL_{la+1} \Rightarrow \forall po \in policiesOfCondition(oe)$
$\forall pr \in propertiesOfOpInv(oi).pr \in visibleProperties(po)$

$\overset{\substack{(3.89),(3.127),\\(3.139),(4.204),(4.102),\\(3.89),(4.120),(4.101),\\(4.105),(4.201),\text{I.H.}}}{\Longleftrightarrow}$ $((oe, oi) \in OEL_{la+1} \Rightarrow \forall po \in policiesOfCondition(oe)$
$\forall pr \in propertiesOfOpInv(oi).pr \in visibleProperties(po))$
$\wedge (\forall po \in policiesOfCondition(oe'')$
$\forall pr \in propertiesOfOpInv(oi'').pr \in visibleProperties(po))$

$\Longleftrightarrow$ $((oe, oi) \in OEL_{la+1} \Rightarrow \forall po \in policiesOfCondition(oe)$
$\forall pr \in propertiesOfOpInv(oi).pr \in visibleProperties(po))$
$\wedge ((oe, oi) = (oe'', oi'') \Rightarrow \forall po \in policiesOfCondition(oe)$
$\forall pr \in propertiesOfOpInv(oi).pr \in visibleProperties(po))$

$\Longleftrightarrow$ $(oe, oi) \in OEL_{la+1} \cup (oe'', oi'') \Rightarrow \forall po \in policiesOfCondition(oe)$
$\forall pr \in propertiesOfOpInv(oi).pr \in visibleProperties(po)$

$\overset{(4.203)}{\Longleftrightarrow}$ $(oe, oi) \in OEL'_{la+1} \Rightarrow \forall po \in policiesOfCondition(oe)$
$\forall pr \in propertiesOfOpInv(oi).pr \in visibleProperties(po)$

$\overset{(3.158)}{\Longleftrightarrow}$ (3.158) holds for LM'_{la+1}

$$(3.159) \text{ holds for } LM'_{la}$$

$$\overset{(3.159)}{\Longleftrightarrow} \quad (oe, oi) \in OEL'_{la}$$
$$\Rightarrow layersOfCondition(oe) \subseteq layersOfOpInv(oi)$$

$$\overset{(4.201)}{\Longleftrightarrow} \quad (oe, oi) \in OEL_{la} \cup (oe', oi')$$
$$\Rightarrow layersOfCondition(oe) \subseteq layersOfOpInv(oi)$$

$$\Longrightarrow \quad (oe, oi) \in OEL_{la}$$
$$\Rightarrow layersOfCondition(oe) \subseteq layersOfOpInv(oi)$$

$$\overset{\text{I.H.}}{\Longrightarrow} \quad (oe, oi) \in OEL_{la+1}$$
$$\Rightarrow layersOfCondition(oe) \subseteq layersOfOpInv(oi)$$

$$\overset{\substack{(4.204),(4.95),(4.99),\\(3.129),(3.128)}}{\Longleftrightarrow} \quad ((oe, oi) \in OEL_{la+1}$$
$$\Rightarrow layersOfCondition(oe) \subseteq layersOfOpInv(oi))$$
$$\wedge \, layersOfCondition(oe'') \subseteq layersOfOpInv(oi'')$$

$$\overset{}{\Longleftrightarrow} \quad ((oe, oi) \in OEL_{la+1}$$
$$\Rightarrow layersOfCondition(oe) \subseteq layersOfOpInv(oi))$$
$$\wedge \, ((oe, oi) = (oe'', oi''))$$
$$\Rightarrow layersOfCondition(oe) \subseteq layersOfOpInv(oi))$$

$$\overset{}{\Longleftrightarrow} \quad (oe, oi) \in OEL_{la+1} \cup (oe'', oi'')$$
$$\Rightarrow layersOfCondition(oe) \subseteq layersOfOpInv(oi)$$

$$\overset{(4.203)}{\Longleftrightarrow} \quad (oe, oi) \in OEL'_{la+1}$$
$$\Rightarrow layersOfCondition(oe) \subseteq layersOfOpInv(oi)$$

$$\overset{(3.159)}{\Longleftrightarrow} \quad (3.159) \text{ holds for } LM'_{la+1}$$

B.9 Extension with an Action Link

(3.163) holds for LM'_{la}

$\overset{(3.163)}{\Longleftrightarrow}$ $(ac, oi) \in AL'_{la} \Rightarrow ac \in Ac \wedge oi \in OIL1$

$\overset{(4.205)}{\Longleftrightarrow}$ $(ac, oi) \in AL_{la} \cup (ac', oi') \Rightarrow ac \in Ac \wedge oi \in OIL1$

$\Longrightarrow$ $(ac, oi) \in AL_{la} \Rightarrow ac \in Ac \wedge oi \in OIL1$

$\overset{\text{I.H.}}{\Longrightarrow}$ $(ac, oi) \in AL_{la+1} \Rightarrow ac \in Ac \wedge oi \in OIL1$

$\overset{(4.206),(4.89),(4.93)}{\Longleftrightarrow}$ $((ac, oi) \in AL_{la+1} \Rightarrow ac \in Ac \wedge oi \in OIL1)$
         $\wedge \, ac'' \in actionsOfLayer(la+1)$

$\overset{(3.112),(3.108)}{\Longrightarrow}$ $((ac, oi) \in AL_{la+1} \Rightarrow ac \in Ac \wedge oi \in OIL1)$
         $\wedge \, ac'' \in Ac$

$\overset{(4.208),(4.95),(4.99)}{\Longleftrightarrow}$ $((ac, oi) \in AL_{la+1} \Rightarrow ac \in Ac \wedge oi \in OIL1)$
         $\wedge \, ac'' \in Ac$
         $\wedge \, oi'' \in opInvsOfLayer(la+1)$

$\overset{(3.129)}{\Longrightarrow}$ $((ac, oi) \in AL_{la+1} \Rightarrow ac \in Ac \wedge oi \in OIL1)$
         $\wedge \, ac'' \in Ac$
         $\wedge \, oi'' \in OIL1$

$\Longleftrightarrow$ $((ac, oi) \in AL_{la+1} \Rightarrow ac \in Ac \wedge oi \in OIL1)$
         $\wedge \, ((ac, oi) = (ac'', oi'') \Rightarrow ac \in Ac \wedge oi \in OIL1)$

$\Longleftrightarrow$ $(ac, oi) \in AL_{la+1} \cup (ac'', oi'') \Rightarrow ac \in Ac \wedge oi \in OIL1$

$\overset{(4.207)}{\Longleftrightarrow}$ $(ac, oi) \in AL'_{la+1} \Rightarrow ac \in Ac \wedge oi \in OIL1$

$\overset{(3.163)}{\Longleftrightarrow}$ (3.163) holds for LM'_{la+1}

$$\text{(3.164) holds for } LM'_{la}$$

$$\overset{(3.164)}{\Longleftrightarrow} \quad (ac, oi_1) \in AL'_{la}$$
$$\wedge (ac, oi_2) \in AL'_{la}$$
$$\Rightarrow oi_1 = oi_2$$

$$\overset{(4.205)}{\Longleftrightarrow} \quad (ac, oi_1) \in AL_{la} \cup (ac', oi')$$
$$\wedge (ac, oi_2) \in AL_{la} \cup (ac', oi')$$
$$\Rightarrow oi_1 = oi_2$$

$$\Longrightarrow \quad (ac, oi_1) \in AL_{la} \wedge (ac, oi_2) \in AL_{la}$$
$$\Rightarrow oi_1 = oi_2$$

$$\overset{\text{I.H.}}{\Longrightarrow} \quad (ac, oi_1) \in AL_{la+1} (ac, oi_2) \in AL_{la+1}$$
$$\Rightarrow oi_1 = oi_2$$

$$\overset{\substack{(4.116),(4.206),(4.91), \\ (4.205),\text{I.H.},(4.207)}}{\Longleftrightarrow} \quad ((ac, oi_1) \in AL_{la+1} \wedge (ac, oi_2) \in AL_{la+1}$$
$$\Rightarrow oi_1 = oi_2)$$
$$\wedge ((ac, oi_1) \in AL_{la+1} \wedge (ac, oi_2) = (ac'', oi'')$$
$$\Rightarrow oi_1 = oi_2)$$
$$\wedge ((ac, oi_2) \in AL_{la+1} \wedge (ac, oi_1) = (ac'', oi'')$$
$$\Rightarrow oi_1 = oi_2)$$

$$\Longleftrightarrow \quad ((ac, oi_1) \in AL_{la+1} \wedge (ac, oi_2) \in AL_{la+1}$$
$$\Rightarrow oi_1 = oi_2)$$
$$\wedge ((ac, oi_1) \in AL_{la+1} \wedge (ac, oi_2) = (ac'', oi'')$$
$$\Rightarrow oi_1 = oi_2)$$
$$\wedge ((ac, oi_1) = (ac'', oi'') \wedge (ac, oi_2) \in AL_{la+1}$$
$$\Rightarrow oi_1 = oi_2)$$
$$\wedge ((ac, oi_1) = (ac'', oi'') \wedge (ac, oi_2) = (ac'', oi'')$$
$$\Rightarrow co_1 = co_2)$$

$$\Longleftrightarrow \quad ((ac, oi_1) \in AL_{la+1} \vee (ac, oi_1) = (ac'', oi''))$$
$$\wedge ((ac, oi_2) \in AL_{la+1} \vee (ac, oi_2) = (ac'', oi''))$$
$$\Rightarrow oi_1 = oi_2$$

$$\Longleftrightarrow \quad (ac, oi_1) \in AL_{la+1} \cup (ac'', oi'')$$
$$\wedge (ac, oi_2) \in AL_{la+1} \cup (ac'', oi'')$$
$$\Rightarrow oi_1 = oi_2$$

$$\overset{(4.207)}{\Longleftrightarrow} \quad (ac, oi_1) \in AL'_{la+1}$$
$$\wedge (ac, oi_2) \in AL'_{la+1}$$
$$\Rightarrow oi_1 = oi_2$$

$$\overset{(3.164)}{\Longleftrightarrow} \quad \text{(3.164) holds for } LM'_{la+1}$$

(3.165) holds for LM'_{la}

$\overset{(3.165)}{\Longleftrightarrow}$ $(ac, oi) \in AL'_{la} \Rightarrow \forall po \in policiesOfAction(ac)$
$\forall pr \in propertiesOfOpInv(oi).pr \in visibleProperties(po)$

$\overset{(4.205)}{\Longleftrightarrow}$ $(ac, oi) \in AL_{la} \cup (ac', oi') \Rightarrow \forall po \in policiesOfAction(ac)$
$\forall pr \in propertiesOfOpInv(oi).pr \in visibleProperties(po)$

$\Longrightarrow$ $(ac, oi) \in AL_{la} \Rightarrow \forall po \in policiesOfAction(ac)$
$\forall pr \in propertiesOfOpInv(oi).pr \in visibleProperties(po)$

$\overset{I.H.}{\Longrightarrow}$ $(ac, oi) \in AL_{la+1} \Rightarrow \forall po \in policiesOfAction(ac)$
$\forall pr \in propertiesOfOpInv(oi).pr \in visibleProperties(po)$

$\overset{\begin{subarray}{c}(3.109),(3.127),\\(3.139),(4.208),(4.102),\\(3.109),(4.120),(4.101),\\(4.105),(4.205),I.H.\end{subarray}}{\Longleftrightarrow}$ $((ac, oi) \in AL_{la+1} \Rightarrow \forall po \in policiesOfAction(ac)$
$\forall pr \in propertiesOfOpInv(oi).pr \in visibleProperties(po))$
$\wedge (\forall po \in policiesOfAction(ac'')$
$\forall pr \in propertiesOfOpInv(oi'').pr \in visibleProperties(po))$

$\Longleftrightarrow$ $((ac, oi) \in AL_{la+1} \Rightarrow \forall po \in policiesOfAction(ac)$
$\forall pr \in propertiesOfOpInv(oi).pr \in visibleProperties(po))$
$\wedge ((ac, oi) = (ac'', oi'') \Rightarrow \forall po \in policiesOfAction(ac)$
$\forall pr \in propertiesOfOpInv(oi).pr \in visibleProperties(po))$

$\Longleftrightarrow$ $(ac, oi) \in AL_{la+1} \cup (ac'', oi'') \Rightarrow \forall po \in policiesOfAction(ac)$
$\forall pr \in propertiesOfOpInv(oi).pr \in visibleProperties(po)$

$\overset{(4.207)}{\Longleftrightarrow}$ $(ac, oi) \in AL'_{la+1} \Rightarrow \forall po \in policiesOfAction(ac)$
$\forall pr \in propertiesOfOpInv(oi).pr \in visibleProperties(po)$

$\overset{(3.165)}{\Longleftrightarrow}$ (3.165) holds for LM'_{la+1}

$$\text{(3.166) holds for } LM'_{la}$$

$$\overset{(3.166)}{\Longleftrightarrow} \quad (ac, oi) \in AL'_{la}$$
$$\Rightarrow layersOfAction(ac) \subseteq layersOfOpInv(oi)$$

$$\overset{(4.205)}{\Longleftrightarrow} \quad (ac, oi) \in AL_{la} \cup (ac', oi')$$
$$\Rightarrow layersOfAction(ac) \subseteq layersOfOpInv(oi)$$

$$\Longrightarrow \quad (ac, oi) \in AL_{la}$$
$$\Rightarrow layersOfAction(ac) \subseteq layersOfOpInv(oi)$$

$$\overset{\text{I.H.}}{\Longrightarrow} \quad (ac, oi) \in AL_{la+1}$$
$$\Rightarrow layersOfAction(ac) \subseteq layersOfOpInv(oi)$$

$$\overset{\substack{(4.208),(4.95),(4.99),\\(3.129),(3.128)}}{\Longleftrightarrow} \quad ((ac, oi) \in AL_{la+1}$$
$$\Rightarrow layersOfAction(ac) \subseteq layersOfOpInv(oi))$$
$$\wedge layersOfAction(ac'') \subseteq layersOfOpInv(oi'')$$

$$\Longleftrightarrow \quad ((ac, oi) \in AL_{la+1}$$
$$\Rightarrow layersOfAction(ac) \subseteq layersOfOpInv(oi))$$
$$\wedge ((ac, oi) = (ac'', oi''))$$
$$\Rightarrow layersOfAction(ac) \subseteq layersOfOpInv(oi))$$

$$\Longleftrightarrow \quad (ac, oi) \in AL_{la+1} \cup (ac'', oi'')$$
$$\Rightarrow layersOfAction(ac) \subseteq layersOfOpInv(oi)$$

$$\overset{(4.207)}{\Longleftrightarrow} \quad (ac, oi) \in AL'_{la+1}$$
$$\Rightarrow layersOfAction(ac) \subseteq layersOfOpInv(oi)$$

$$\overset{(3.166)}{\Longleftrightarrow} \quad \text{(3.166) holds for } LM'_{la+1}$$

B.10 Extension with an Operation Invoking Link

Extension with a Link to an Operation

(3.120) holds for LM'_{la}

$$\overset{(3.120)}{\Longleftrightarrow} \quad (oi, op) \in OIL2'_{la} \Rightarrow oi \in OIL1 \land op \in Op$$

$$\overset{(4.213)}{\Longleftrightarrow} \quad (oi, op) \in OIL2_{la} \cup (oi', op') \Rightarrow oi \in OIL1 \land op \in Op$$

$$\Longrightarrow \quad (oi, op) \in OIL2_{la} \Rightarrow oi \in OIL1 \land op \in Op$$

$$\overset{\text{I.H.}}{\Longrightarrow} \quad (oi, op) \in OIL2_{la+1} \Rightarrow oi \in OIL1 \land op \in Op$$

$$\overset{(4.215),(4.95),(4.99)}{\Longleftrightarrow} \quad ((oi, op) \in OIL2_{la+1} \Rightarrow oi \in OIL1 \land op \in Op)$$
$$\land oi'' \in opInvsOfLayer(la + 1)$$

$$\overset{(3.129),(3.116)}{\Longrightarrow} \quad ((oi, op) \in OIL2_{la+1} \Rightarrow oi \in OIL1 \land op \in Op)$$
$$\land oi'' \in OIL1$$

$$\overset{\substack{(4.216),(4.40),(4.16),\\(4.19),(4.156)}}{\Longleftrightarrow} \quad ((oi, op) \in OIL2_{la+1} \Rightarrow oi \in OIL1 \land op \in Op)$$
$$\land oi'' \in OIL1$$
$$\land op''_i \in operationsOfLayer(la + 1)$$

$$\overset{(3.38),(3.36)}{\Longrightarrow} \quad ((oi, op) \in OIL2_{la+1} \Rightarrow oi \in OIL1 \land op \in Op)$$
$$\land oi'' \in OIL1$$
$$\land op''_i \in Op$$

$$\Longleftrightarrow \quad ((oi, op) \in OIL2_{la+1} \Rightarrow oi \in OIL1 \land op \in Op)$$
$$\land ((oi, op) = (oi'', op''_i) \Rightarrow oi \in OIL1 \land op \in Op)$$

$$\Longleftrightarrow \quad (oi, op) \in OIL2_{la+1} \cup (oi'', op''_i) \Rightarrow oi \in OIL1 \land op \in Op$$

$$\overset{(4.219)}{\Longleftrightarrow} \quad (oi, op) \in OIL2'_{la+1} \Rightarrow oi \in OIL1 \land op \in Op$$

$$\overset{(3.120)}{\Longleftrightarrow} \quad (3.120) \text{ holds for } LM'_{la+1}$$

$$(3.121) \text{ holds for } LM'_{la}$$

$$\overset{(3.121)}{\Longleftrightarrow} \quad (oi, op_1) \in OIL2'_{la}$$
$$\wedge (oi, op_2) \in OIL2'_{la}$$
$$\Rightarrow op_1 = op_2$$

$$\overset{(4.213)}{\Longleftrightarrow} \quad (oi, op_1) \in OIL2_{la} \cup (oi', op')$$
$$\wedge (oi, op_2) \in OIL2_{la} \cup (oi', op')$$
$$\Rightarrow op_1 = op_2$$

$$\Longrightarrow \quad (oi, op_1) \in OIL2_{la} \wedge (oi, op_2) \in OIL2_{la}$$
$$\Rightarrow op_1 = op_2$$

$$\overset{\text{I.H.}}{\Longrightarrow} \quad (oi, op_1) \in OIL2_{la+1} \wedge (oi, op_2) \in OIL2_{la+1}$$
$$\Rightarrow op_1 = op_2$$

$$\overset{\substack{(4.118),(4.215),(4.97),\\ (4.213),\text{I.H.},(4.219)}}{\Longleftrightarrow} \quad ((oi, op_1) \in OIL2_{la+1} \wedge (oi, op_2) \in OIL2_{la+1}$$
$$\Rightarrow op_1 = op_2)$$
$$\wedge ((oi, op_1) \in OIL2_{la+1} \wedge (oi, op_2) = (oi'', op''_i)$$
$$\Rightarrow op_1 = op_2)$$
$$\wedge ((oi, op_2) \in OIL2_{la+1} \wedge (oi, op_1) = (oi'', op''_i)$$
$$\Rightarrow op_1 = op_2)$$

$$\Longleftrightarrow \quad ((oi, op_1) \in OIL2_{la+1} \wedge (oi, op_2) \in OIL2_{la+1}$$
$$\Rightarrow op_1 = op_2)$$
$$\wedge ((oi, op_1) \in OIL2_{la+1} \wedge (oi, op_2) = (oi'', op''_i)$$
$$\Rightarrow op_1 = op_2)$$
$$\wedge ((oi, op_1) = (oi'', op''_i) \wedge (oi, op_2) \in OIL2_{la+1}$$
$$\Rightarrow op_1 = op_2)$$
$$\wedge ((oi, op_1) = (oi'', op''_i) \wedge (oi, op_2) = (oi'', op''_i)$$
$$\Rightarrow op_1 = op_2)$$

$$\Longleftrightarrow \quad ((oi, op_1) \in OIL2_{la+1} \vee (oi, op_1) = (oi'', op''_i))$$
$$\wedge ((oi, op_2) \in OIL2_{la+1} \vee (oi, op_2) = (oi'', op''_i))$$
$$\Rightarrow op_1 = op_2$$

$$\Longleftrightarrow \quad (oi, op_1) \in OIL2_{la+1} \cup (oi'', op''_i)$$
$$\wedge (oi, op_2) \in OIL2_{la+1} \cup (oi'', op''_i)$$
$$\Rightarrow op_1 = op_2$$

$$\overset{(4.219)}{\Longleftrightarrow} \quad (oi, op_1) \in OIL2'_{la+1}$$
$$\wedge (oi, op_2) \in OIL2'_{la+1}$$
$$\Rightarrow op_1 = op_2$$

$$\overset{(3.121)}{\Longleftrightarrow} \quad (3.121) \text{ holds for } LM'_{la+1}$$

(3.122) holds for LM'_{la}

$\overset{(3.122)}{\Longleftrightarrow}$ $(oi, pa, arg) \in OIL3'_{la} \Rightarrow oi \in OIL1 \wedge pa \in Pa$
$\wedge (arg \in Id_{Pr} \Rightarrow arg \in Pr) \wedge (arg \in Id_{OI} \Rightarrow arg \in OIL1)$

$\overset{(4.214)}{\Longleftrightarrow}$ $(oi, pa, arg) \in OIL3_{la} \Rightarrow oi \in OIL1 \wedge pa \in Pa$
$\wedge (arg \in Id_{Pr} \Rightarrow arg \in Pr) \wedge (arg \in Id_{OI} \Rightarrow arg \in OIL1)$

$\overset{\text{I.H.}}{\Longrightarrow}$ $(oi, pa, arg) \in OIL3_{la+1} \Rightarrow oi \in OIL1 \wedge pa \in Pa$
$\wedge (arg \in Id_{Pr} \Rightarrow arg \in Pr) \wedge (arg \in Id_{OI} \Rightarrow arg \in OIL1)$

$\overset{(4.225),(4.95),(4.99)}{\Longleftrightarrow}$ $((oi, pa, arg) \in OIL3_{la+1} \Rightarrow oi \in OIL1 \wedge pa \in Pa$
$\wedge (arg \in Id_{Pr} \Rightarrow arg \in Pr) \wedge (arg \in Id_{OI} \Rightarrow arg \in OIL1))$
$\wedge oi'' \in opInvsOfLayer(la + 1)$

$\overset{(3.129),(3.116)}{\Longleftrightarrow}$ $((oi, pa, arg) \in OIL3_{la+1} \Rightarrow oi \in OIL1 \wedge pa \in Pa$
$\wedge (arg \in Id_{Pr} \Rightarrow arg \in Pr) \wedge (arg \in Id_{OI} \Rightarrow arg \in OIL1))$
$\wedge oi'' \in OIL1$

$\overset{\substack{(4.226),(4.42),(4.22),\\(4.25),(4.230)}}{\Longleftrightarrow}$ $((oi, pa, arg) \in OIL3_{la+1} \Rightarrow oi \in OIL1 \wedge pa \in Pa$
$\wedge (arg \in Id_{Pr} \Rightarrow arg \in Pr) \wedge (arg \in Id_{OI} \rightarrow arg \in OIL1))$
$\wedge oi'' \in OIL1 \wedge pa''_i \, parametersOfLayer(la + 1)$

$\overset{(4.225),(4.95),(4.99)}{\Longleftrightarrow}$ $((oi, pa, arg) \in OIL3_{la+1} \Rightarrow oi \in OIL1 \wedge pa \in Pa$
$\wedge (arg \in Id_{Pr} \Rightarrow arg \in Pr) \wedge (arg \in Id_{OI} \Rightarrow arg \in OIL1))$
$\wedge oi'' \in OIL1 \wedge pa''_i \in Pa$

$\overset{(4.221),(4.67),(3.11)}{\Longleftrightarrow}$ $((oi, pa, arg) \in OIL3_{la+1} \Rightarrow oi \in OIL1 \wedge pa \in Pa$
$\wedge (arg \in Id_{Pr} \Rightarrow arg \in Pr) \wedge (arg \in Id_{OI} \Rightarrow arg \in OIL1))$
$\wedge oi'' \in OIL1 \wedge pa''_i \in Pa$
$\wedge (lit'' \in Id_{Pr} \Rightarrow lit'' \in Pr) \wedge (lit'' \in Id_{OI} \Rightarrow lit'' \in OIL1)$

$\Longleftrightarrow$ $((oi, pa, arg) \in OIL3_{la+1} \Rightarrow oi \in OIL1 \wedge pa \in Pa$
$\wedge (arg \in Id_{Pr} \Rightarrow arg \in Pr) \wedge (arg \in Id_{OI} \Rightarrow arg \in OIL1))$
$\wedge ((oi, pa, arg) = (oi'', pa''_i, lit'') \Rightarrow oi \in OIL1 \wedge pa \in Pa$
$\wedge (arg \in Id_{Pr} \Rightarrow arg \in Pr) \wedge (arg \in Id_{OI} \Rightarrow arg \in OIL1))$

$\Longleftrightarrow$ $(oi, pa, arg) \in OIL3_{la+1} \cup (oi'', pa''_i, arg'') \Rightarrow oi \in OIL1 \wedge pa \in Pa$
$\wedge (arg \in Id_{Pr} \Rightarrow arg \in Pr) \wedge (arg \in Id_{OI} \Rightarrow arg \in OIL1)$

$\overset{(4.220)}{\Longleftrightarrow}$ $(oi, pa, arg) \in OIL3'_{la+1} \Rightarrow oi \in OIL1 \wedge pa \in Pa$
$\wedge (arg \in Id_{Pr} \Rightarrow arg \in Pr) \wedge (arg \in Id_{OI} \Rightarrow arg \in OIL1)$

$\overset{(3.122)}{\Longleftrightarrow}$ (3.122) holds for LM'_{la+1}

$$(3.123) \text{ holds for } PM'_{la}$$

$$\overset{(3.123)}{\Longleftrightarrow} \quad (oi, pa, arg) \in OIL3'_{la} \Rightarrow \exists op.(oi, op) \in OIL2'_{la}$$
$$\land\ pa \in parametersOfOperation(op)$$

$$\overset{(4.214),(4.213)}{\Longleftrightarrow} \quad (oi, pa, arg) \in OIL3_{la} \Rightarrow \exists op.(oi, op) \in OIL2_{la} \cup (oi', op')$$
$$\land\ pa \in parametersOfOperation(op)$$

$$\overset{\text{I.H.}}{\Longleftrightarrow} \quad (oi, pa, arg) \in OIL3_{la} \Rightarrow \exists op.(oi, op) \in OIL2_{la}$$
$$\land\ pa \in parametersOfOperation(op)$$

$$\overset{\text{I.H.}}{\Longrightarrow} \quad (oi, pa, arg) \in OIL3_{la+1} \Rightarrow \exists op.(oi, op) \in OIL2_{la+1}$$
$$\land\ pa \in parametersOfOperation(op)$$

$$\overset{(4.220),(3.44),(3.45)}{\Longrightarrow} \quad ((oi, pa, arg) \in OIL3_{la+1} \Rightarrow \exists op.(oi, op) \in OIL2_{la+1}$$
$$\land\ pa \in parametersOfOperation(op))$$
$$\land\ \exists op.(oi'', op) \in OIL2_{la+1}$$
$$\land\ pa''_i \in parametersOfOperation(op)$$

$$\overset{}{\Longleftrightarrow} \quad ((oi, pa, arg) \in OIL3_{la+1} \Rightarrow \exists op.(oi, op) \in OIL2_{la+1}$$
$$\land\ pa \in parametersOfOperation(op))$$
$$\land\ ((oi, pa, arg) = (oi'', pa''_i, lit'') \Rightarrow \exists op.(oi, op) \in OIL2_{la+1}$$
$$\land\ pa \in parametersOfOperation(op))$$

$$\overset{}{\Longleftrightarrow} \quad (oi, pa, arg) \in OIL3_{la+1} \cup (oi'', pa''_i, arg'') \Rightarrow \exists op.(oi, op) \in OIL2_{la+1}$$
$$\land\ pa \in parametersOfOperation(op)$$

$$\overset{(4.220),(4.219)}{\Longleftrightarrow} \quad (oi, pa, arg) \in OIL3'_{la+1} \Rightarrow \exists op.(oi, op) \in OIL2'_{la+1}$$
$$\land\ pa \in parametersOfOperation(op)$$

$$\overset{(3.123)}{\Longleftrightarrow} \quad (3.123) \text{ holds for } PM'_{la+1}$$

Extension with a Link to a Parameter and Argument

We prove the semantical correctness for the case of a literal as argument of the operation invoking link, according to section 4.2.3, case 12a. The other cases 12b and 12c can be proved accordingly.

$$(3.120) \text{ holds for } LM'_{la}$$

$$\overset{(3.120)}{\Longleftrightarrow} \quad (oi, op) \in OIL2'_{la} \Rightarrow oi \in OIL1 \wedge op \in Op$$

$$\overset{(4.223)}{\Longleftrightarrow} \quad (oi, op) \in OIL2_{la} \Rightarrow oi \in OIL1 \wedge op \in Op$$

$$\overset{\text{I.H.}}{\Longrightarrow} \quad (oi, op) \in OIL2_{la+1} \Rightarrow oi \in OIL1 \wedge op \in Op$$

$$\overset{(4.229)}{\Longleftrightarrow} \quad (oi, op) \in OIL2'_{la+1} \Rightarrow oi \in OIL1 \wedge op \in Op$$

$$\overset{(3.120)}{\Longleftrightarrow} \quad (3.120) \text{ holds for } LM'_{la+1}$$

(3.121) holds for LM'_{la}

$\overset{(3.121)}{\Longleftrightarrow}$ $(oi, op_1) \in OIL2'_{la}$
$\land\, (oi, op_2) \in OIL2'_{la}$
$\Rightarrow op_1 = op_2$

$\overset{(4.223)}{\Longleftrightarrow}$ $(oi, op_1) \in OIL2_{la}$
$\land\, (oi, op_2) \in OIL2_{la} \cup (oi', op')$
$\Rightarrow op_1 = op_2$

$\overset{\text{I.H.}}{\Longrightarrow}$ $(oi, op_1) \in OIL2_{la+1} \land (oi, op_2) \in OIL2_{la+1}$
$\Rightarrow op_1 = op_2$

$\overset{(4.229)}{\Longleftrightarrow}$ $(oi, op_1) \in OIL2'_{la+1}$
$\land\, (oi, op_2) \in OIL2'_{la+1}$
$\Rightarrow op_1 = op_2$

$\overset{(3.121)}{\Longleftrightarrow}$ (3.121) holds for LM'_{la+1}

(3.122) holds for LM'_{la}

$\overset{(3.122)}{\Longleftrightarrow}$ $(oi, pa, arg) \in OIL3'_{la} \Rightarrow oi \in OIL1 \land pa \in Pa$
$\land (arg \in Id_{Pr} \Rightarrow arg \in Pr) \land (arg \in Id_{OI} \Rightarrow arg \in OIL1)$

$\overset{(4.224)}{\Longleftrightarrow}$ $(oi, pa, arg) \in OIL3_{la} \cup (oi', pa', arg') \Rightarrow oi \in OIL1 \land pa \in Pa$
$\land (arg \in Id_{Pr} \Rightarrow arg \in Pr) \land (arg \in Id_{OI} \Rightarrow arg \in OIL1)$

$\Longrightarrow$ $(oi, pa, arg) \in OIL3_{la} \Rightarrow oi \in OIL1 \land pa \in Pa$
$\land (arg \in Id_{Pr} \Rightarrow arg \in Pr) \land (arg \in Id_{OI} \Rightarrow arg \in OIL1)$

$\overset{I.H.}{\Longrightarrow}$ $(oi, pa, arg) \in OIL3_{la+1} \Rightarrow oi \in OIL1 \land pa \in Pa$
$\land (arg \in Id_{Pr} \Rightarrow arg \in Pr) \land (arg \in Id_{OI} \Rightarrow arg \in OIL1)$

$\overset{(4.225),(4.95),(4.99)}{\Longleftrightarrow}$ $((oi, pa, arg) \in OIL3_{la+1} \Rightarrow oi \in OIL1 \land pa \in Pa$
$\land (arg \in Id_{Pr} \Rightarrow arg \in Pr) \land (arg \in Id_{OI} \Rightarrow arg \in OIL1))$
$\land oi'' \in opInvsOfLayer(la+1)$

$\overset{(3.129),(3.116)}{\Longleftrightarrow}$ $((oi, pa, arg) \in OIL3_{la+1} \Rightarrow oi \in OIL1 \land pa \in Pa$
$\land (arg \in Id_{Pr} \Rightarrow arg \in Pr) \land (arg \in Id_{OI} \Rightarrow arg \in OIL1))$
$\land oi'' \in OIL1$

$\overset{(4.226),(4.42),(4.22),}{\underset{(4.25),(4.230)}{\Longleftrightarrow}}$ $((oi, pa, arg) \in OIL3_{la+1} \Rightarrow oi \in OIL1 \land pa \in Pa$
$\land (arg \in Id_{Pr} \Rightarrow arg \in Pr) \land (arg \in Id_{OI} \Rightarrow arg \in OIL1))$
$\land oi'' \in OIL1 \land pa''_i parametersOfLayer(la+1)$

$\overset{(4.225),(4.95),(4.99)}{\Longleftrightarrow}$ $((oi, pa, arg) \in OIL3_{la+1} \Rightarrow oi \in OIL1 \land pa \in Pa$
$\land (arg \in Id_{Pr} \Rightarrow arg \in Pr) \land (arg \in Id_{OI} \Rightarrow arg \in OIL1))$
$\land oi'' \in OIL1 \land pa''_i \in Pa$

$\overset{(4.231),(4.27),(3.11)}{\Longleftrightarrow}$ $((oi, pa, arg) \in OIL3_{la+1} \Rightarrow oi \in OIL1 \land pa \in Pa$
$\land (arg \in Id_{Pr} \Rightarrow arg \in Pr) \land (arg \in Id_{OI} \Rightarrow arg \in OIL1))$
$\land oi'' \in OIL1 \land pa''_i \in Pa$
$\land (arg'' \in Id_{Pr} \Rightarrow arg'' \in Pr) \land (arg'' \in Id_{OI} \Rightarrow arg'' \in OIL1)$

$\Longleftrightarrow$ $((oi, pa, arg) \in OIL3_{la+1} \Rightarrow oi \in OIL1 \land pa \in Pa$
$\land (arg \in Id_{Pr} \Rightarrow arg \in Pr) \land (arg \in Id_{OI} \Rightarrow arg \in OIL1))$
$\land ((oi, pa, arg) = (oi'', pa''_i, arg'') \Rightarrow oi \in OIL1 \land pa \in Pa$
$\land (arg \in Id_{Pr} \Rightarrow arg \in Pr) \land (arg \in Id_{OI} \Rightarrow arg \in OIL1))$

$\Longleftrightarrow$ $(oi, pa, arg) \in OIL3_{la+1} \cup (oi'', pa''_i, arg'') \Rightarrow oi \in OIL1 \land pa \in Pa$
$\land (arg \in Id_{Pr} \Rightarrow arg \in Pr) \land (arg \in Id_{OI} \Rightarrow arg \in OIL1)$

$\overset{(4.230)}{\Longleftrightarrow}$ $(oi, pa, arg) \in OIL3'_{la+1} \Rightarrow oi \in OIL1 \land pa \in Pa$
$\land (arg \in Id_{Pr} \Rightarrow arg \in Pr) \land (arg \in Id_{OI} \Rightarrow arg \in OIL1)$

$\overset{(3.122)}{\Longleftrightarrow}$ (3.122) holds for LM'_{la+1}

(3.123) holds for PM'_{la}

$\overset{(3.123)}{\Longleftrightarrow}$ $(oi, pa, arg) \in OIL3'_{la} \Rightarrow \exists op.(oi, op) \in OIL2'_{la}$
$\wedge\, pa \in parametersOfOperation(op)$

$\overset{(4.224),(4.223)}{\Longleftrightarrow}$ $(oi, pa, arg) \in OIL3_{la} \cup (oi', pa', arg') \Rightarrow \exists op.(oi, op) \in OIL2_{la}$
$\wedge\, pa \in parametersOfOperation(op)$

$\Longrightarrow$ $(oi, pa, arg) \in OIL3_{la} \Rightarrow \exists op.(oi, op) \in OIL2_{la}$
$\wedge\, pa \in parametersOfOperation(op)$

$\overset{I.H.}{\Longrightarrow}$ $(oi, pa, arg) \in OIL3_{la+1} \Rightarrow \exists op.(oi, op) \in OIL2_{la+1}$
$\wedge\, pa \in parametersOfOperation(op)$

$\overset{(4.230),(3.44),(3.45)}{\Longrightarrow}$ $((oi, pa, arg) \in OIL3_{la+1} \Rightarrow \exists op.(oi, op) \in OIL2_{la+1}$
$\wedge\, pa \in parametersOfOperation(op))$
$\wedge\, \exists op.(oi'', op) \in OIL2_{la+1}$
$\wedge\, pa''_i \in parametersOfOperation(op)$

$\Longleftrightarrow$ $((oi, pa, arg) \in OIL3_{la+1} \Rightarrow \exists op.(oi, op) \in OIL2_{la+1}$
$\wedge\, pa \in parametersOfOperation(op))$
$\wedge\, ((oi, pa, arg) = (oi'', pa''_i, arg'') \Rightarrow \exists op.(oi, op) \in OIL2_{la+1}$
$\wedge\, pa \in parametersOfOperation(op))$

$\Longleftrightarrow$ $(oi, pa, arg) \in OIL3_{la+1} \cup (oi'', pa''_i, arg'') \Rightarrow \exists op.(oi, op) \in OIL2_{la+1}$
$\wedge\, pa \in parametersOfOperation(op)$

$\overset{(4.230),(4.229)}{\Longleftrightarrow}$ $(oi, pa, arg) \in OIL3'_{la+1} \Rightarrow \exists op.(oi, op) \in OIL2'_{la+1}$
$\wedge\, pa \in parametersOfOperation(op)$

$\overset{(3.123)}{\Longleftrightarrow}$ (3.123) holds for PM'_{la+1}

Bibliography

[1] Information and Communication Technology (ICT) Survey. Survey, U.S. Census Bureau, March 2012.

[2] Information and Communication Technology (ICT) Survey. Video transcript, Intel Corporation, 2005.

[3] Wu-chun Feng. Making a Case for Efficient Supercomputing. *Queue*, 1(7):54–64, October 2003.

[4] Chip Walter. Kryder's Law. *Scientific American*, 293(2):32–33, August 2005.

[5] Gordon E. Moore. Cramming more components onto integrated circuits. *Electronics*, 38(8):114–117, April 1965.

[6] Victor V. Zhirnov, Ralph K. Cavin, III, James A. Hutchby, and George I. Bourianoff. Limits to Binary Logic Switch Scaling – A Gedanken Model. *Proceedings of the IEEE*, 91(11):1934–1939, November 2003.

[7] Manek Dubash. Moore's Law is dead, says Gordon Moore. April 2005.

[8] Patrick P. Gelsinger. From Peta FLOPS to Milli Watts, April 2008. Intel Developers Forum Opening Keynote.

[9] Lawrence M. Krauss and Glenn D. Starkman. Universal Limits on Computation. *ArXiv Astrophysics e-prints*, April 2008.

[10] Frederick P. Brooks. *The Mythical Man-Month: Essays on Software Engineering*. Addison-Wesley, August 1995.

[11] A. G. Ganek and T. A. Corbi. The dawning of the autonomic computing era. *Systems*, 42(1):5–18, January 2003.

[12] Frank E. Gillett, Charles Rutstein, Galen Schreck, Christian Buss, and Heather Liddell. Organic IT. Report, Forrester Research, April 2002.

[13] Markus Völter and Thomas Stahl. *Model-Driven Software Development: Technology, Engineering, Management*. Wiley, May 2006.

[14] Jean Bézivin. On the unification power of models. *Software and Systems Modeling*, 4(2):171–188, May 2005.

[15] MDA Guide Version 1.0.1. Specification omg/2003-06-01, Object Management Group, June 2003.

[16] Jack Greenfield, Keith Short, Steve Cook, and Stuart Kent. *Software Factories: Assembling Applications with Patterns, Models, Frameworks, and Tools*. Wiley, August 2004.

[17] Dave Thomas and Brian M. Barry. Model Driven Development – The Case for Domain Oriented Programming. In *Companion of the 18th Conference on Object-oriented Programming, Systems, Languages, and Applications (OOPSLA)*, pages 2–7. Association for Computing Machinery, October 2003.

[18] Nicodemos Damianou, Naranker Dulay, Emil Lupu, and Morris Sloman. The Ponder Policy Specification Language. In *2nd International Workshop on Policies for Distributed Systems and Networks (POLICY)*, pages 18–38. Springer LNCS, January 2001.

[19] John C. Strassner. *Policy-Based Network Management: Solutions for the Next Generation*. Morgan Kaufman, September 2003.

[20] Jeffrey O. Kephart and David M. Chess. The Vision of Autonomic Computing. *Computer*, 36(1):41–50, January 2003.

[21] An architectural blueprint for autonomic computing. White paper, IBM Corporation, June 2006.

[22] Building an adaptive enterprise – Linking business and IT. White paper, Hewlet-Packard Corporation, May 2003.

[23] Dynamic Systems 2007 – Get Started With Dynamic Systems Technology Today. White paper, Microsoft, May 2007.

[24] N1 Grid Technology – Just In Time Computing. White paper, Sun Microsystems Corporation, 2004.

[25] Organic Computing – Computer- und Systemarchitektur im Jahr 2010. Position paper, Gesellschaft für Informatik, Informationstechnische Gesellschaft im VDE, 2003.

[26] IT Projects: Experience Certainty. Independent market research report, Dynamic Markets Limited, August 2007.

[27] CHAOS Report 2007: The Laws of CHAOS. Report, The Standish Group, 2007.

[28] CHAOS Report 1994. Report, The Standish Group, 1994.

[29] Dave Patterson. A New Focus for a New Century: Availability and Maintainability >> Performance, January 2002. USENIX Conference on File and Storage Technologies (FAST) Keynote.

[30] Aaron B. Brown and David A. Patterson. To Err is Human. In *1st Workshop on Evaluating and Architecting System dependabilitY (EASY)*, July 2001.

[31] David Patterson, Aaron Brown, Pete Broadwell, George Candea, Mike Chen, James Cutler, Patricia Enriquez, Armando Fox, Emre Kiciman, Matthew Merzbacher, David Oppenheimer, Naveen Sastry, William Tetzlaff, Jonathan Traupman, and Noah Treuhaft. Recovery-Oriented Computing (ROC): Motivation, Definition, Techniques, and Case Studies. Computer science technical report UCB//CSD-02-1175, U.C. Berkeley, March 2002.

[32] Seppo Hämäläinen, Henning Sanneck, and Cinzia Sartori, editors. *LTE Self-Organising Networks (SON): Network Management Automation for Operational Efficiency*. Wiley, January 2012.

[33] M. M. Lehman, J. F. Ramil, P. D. Wernick, D. E. Perry, and W. M. Turski. Metrics and Laws of Software Evolution – The Nineties View. In *4th International Symposium on Software Metrics (METRICS)*, pages 20–32. IEEE Computer Society, November 1997.

[34] L. A. Belady and M. M. Lehman. A model of large program development. *Systems*, 15(3):225–252, September 1976.

[35] Wladyslaw M. Turski. The Reference Model for Smooth Growth of Software Systems Revisited. *Software Engineering*, 28(8):814–815, August 2002.

[36] Stefan Koch. Software evolution in open source projects – a large-scale investigation. *Software Maintenance and Evolution: Research and Practice*, 19(6):361–382, 2007.

[37] Morris Sloman. Policy Driven Management for Distributed Systems. *Network and Systems Management*, 2(4):333–360, 1994.

[38] Raphael Romeikat and Bernhard Bauer. Specification and Refinement of Domain-Specific ECA Policies. In *4th International Workshop on Domain-specific Engineering (DsE@CAiSE)*, pages 197–206. Springer LNBIP, June 2011.

[39] Raphael Romeikat, Bernhard Bauer, and Henning Sanneck. Modeling of Domain-Specific ECA Policies. In *23rd International Conference on Software Engineering and Knowledge Engineering (SEKE)*, pages 52–58. Knowledge Systems Institute, July 2011.

[40] Raphael Romeikat and Bernhard Bauer. Formal Specification of Domain-Specific ECA Policy Models. In *5th International Conference on Theoretical Aspects of Software Engineering (TASE)*, pages 209–212. IEEE Computer Society, August 2011.

[41] Raphael Romeikat, Stephan Roser, Pascal Müllender, and Bernhard Bauer. Translation of QVT Relations into QVT Operational Mappings. In *International Conference on Model Transformation (ICMT)*, pages 137–151. Springer LNCS, July 2008.

[42] Raphael Romeikat, Markus Sinsel, and Bernhard Bauer. Transformation of Graphical ECA Policies into Executable PonderTalk Code. In *3rd International Symposium on Rule Interchange and Applications (RuleML)*, pages 193–207. Springer LNCS, November 2009.

[43] Raphael Romeikat, Bernhard Bauer, and Henning Sanneck. Automated Refinement of ECA Policies for Network Management. In *17th Asia-Pacific Conference on Communications (APCC)*, pages 439–444. IEEE Communications Society, October 2011.

[44] T. Bandh, G. Carle, H. Sanneck, L. C. Schmelz, R. Romeikat, and B. Bauer. Optimized Network Configuration Parameter Assignment Based on Graph Coloring. In *12th Network Operations and Management Symposium (NOMS)*, pages 40–47. IEEE Communications Society, April 2010.

[45] Raphael Romeikat, Bernhard Bauer, Tobias Bandh, Georg Carle, Henning Sanneck, and Lars Christoph Schmelz. Policy-driven Workflows for Mobile Network Management Automation. In *6th International Wireless Communications and Mobile Computing Conference (IWCMC)*, pages 1111–1115. Association for Computing Machinery, June–July 2010.

[46] Tobias Bandh, Henning Sanneck, and Raphael Romeikat. An Experimental System for SON Function Coordination. In *International Workshop on Self-Organizing Networks (IWSON)*, pages 1–2. IEEE Vehicular Technology Society, May 2011.

[47] Tobias Bandh, Raphael Romeikat, Henning Sanneck, and Haitao Tang. Policy-based Coordination and Management of SON Functions. In *12th International Symposium on Integrated Network Management (IM)*, pages 823–836. IEEE Communications Society, May 2011.

[48] Tobias Bandh, Henning Sanneck, and Raphael Romeikat. Demonstration of an Integrated SON Experimental System for Self-Optimization and SON Coordination. In *2nd International Workshop on Self-Organizing Networks (IWSON)*. IEEE Vehicular Technology Society, August 2012. To appear.

[49] Patent application, Nokia Siemens Networks Oy, June 2011. Not yet disclosed.

[50] R. Romeikat, B. Bauer, H. Sanneck, and C. Schmelz. A Policy-Based System for Network-Wide Configuration Management. In *18th Meeting of the Wireless World Research Forum (WWRF)*. Wireless World Research Forum, June 2007.

[51] Thomas A. Limoncelli, Christine Hogan, and Strata Chalup. *The Practice of System and Network Administration*. Addison-Wesley, July 2007.

[52] Steven Davy, Keara Barrett, Brendan Jennings, and Sven van der Meer. On the use of Policy Based Management for Pervasive m-Government Services (Euro mGov). In *1st European Conference on Mobile Government*, pages 110–121. Mobile Government Consortium International, July 2005.

[53] Mark Burgess. A Site Configuration Engine. *USENIX Computing Systems*, 8(3), 1995.

[54] *Cambridge Business English Dictionary*. Cambridge University, November 2011.

[55] Isaac Asimov. The Evitable Conflict. *Astounding Science Fiction*, 45(4):48–68, June 1950.

[56] Roger Clarke. Asimov's Laws of Robotics: Implications for Information Technology. *Computer*, 26(12):53–61, 1993.

[57] Raouf Boutaba and Issam Aib. Policy-Based Management: A Historical perspective. *Network and Systems Management*, 15(4):447–480, December 2007.

[58] Pierangela Samarati and Sabrina De Capitani di Vimercati. Access Control: Policies, Models, and Mechanisms. In *Revised versions of lectures given during the IFIP WG 1.7 International School on Foundations of Security Analysis and Design on Foundations of Security Analysis and Design (FOSAD)*, pages 137–196. Springer, September 2001.

[59] B. W. Lampson. Dynamic protection structures. In *35th AFIPS Conference (Fall Joint Computer Conference)*, pages 27–38. American Federation of Information Processing Societies, November 1969.

[60] Butler W. Lampson. Protection. In *5th Princeton Conference on Information Sciences and Systems*, pages 437–443. Princeton University, March 1971.

[61] Butler W. Lampson. Protection. *Operating Systems Review*, 8(1):18–24, January 1974.

[62] Jack B. Dennis and Earl C. Van Horn. Programming Semantics for Multiprogrammed Computations. *Communications of the ACM*, 9(3):143–155, March 1966.

[63] R. S. Fabry. Capability-Based Addressing. *Communications of the ACM*, 17(7):403–412, July 1974.

[64] David D. Clark and David R. Wilson. A Comparison of Commercial and Military Computer Security Policies. In *IEEE Symposium on Security and Privacy*, pages 184–194. IEEE Computer Society, April 1987.

[65] D. Elliott Bell and Leonard J. LaPadula. Secure Computer Systems: Mathematical Foundations. Technical Report MTR-2547, The MITRE Corporation, March 1973.

[66] D. E. Bell and L. J. LaPadula. Secure Computer System: Unified Exposition and Multics Interpretation. Technical Report MTR-2997, The MITRE Corporation, March 1976.

[67] David F. Ferraiolo and D. Richard Kuhn. Role-Based Access Controls. In *15th National Computer Security Conference*, pages 554–563. National Institute of Standards and Technology, October 1992.

[68] Ravi S. Sandhu, Edward J. Coyne, Hal L. Feinstein, and Charles E. Youman. Role-Based Access Control Models. *Computer*, 29(2):38–47, February 1996.

[69] Sylvia Osborn, Ravi Sandhu, and Qamar Munawer. Configuring Role-Based Access Control to Enforce Mandatory and Discretionary Access Control Policies. *Transactions on Information and System Security (TISSEC)*, 3(2):85–106, May 2000.

[70] Ravi Sandhu, David Ferraiolo, and Richard Kuhn. The NIST Model for Role-Based Access Control – Towards A Unified Standard. In *5th Workshop on Role-based Access Control (RBAC)*, pages 47–63. Association for Computing Machinery, July 2000.

[71] Role Based Access Control. Standard ANSI INCITS 359-2004, American National Standard for Information Technology, February 2004.

[72] Alan C. O'Connor and Ross J. Loomis. 2010 Economic Analysis of Role-Based Access Control. Final report, National Institute of Standards and Technology, December 2010.

[73] D. C. Robinson and M. S. Sloman. Domains: A New Approach to Distributed System Management. In *Workshop on the Future Trends of Distributed Computing Systems in the 1990s*, pages 154–163. IEEE Computer Society, September 1988.

[74] Morris Sloman. Domains: A New Approach to Distributed System Management. In *Workshop on Operating Systems of the 90s and Beyond*, pages 25–30. Springer LNCS, July 1991.

[75] Mahadev Satyanarayanan. A Survey of Distributed File Systems. *Annual Review Of Computer Science*, 4:73–104, June 1990.

[76] René Wies. Policies in Network and Systems Management – Formal Definition and Architecture. *Network and Systems Management*, 2(1):63–83, March 1994.

[77] K. Ramakrishnan, S. Floyd, and D. Black. The Addition of Explicit Congestion Notification (ECN) to IP. Request for Comments 3168, Internet Engineering Task Force, September 2001.

[78] David D. Clark. Policy Routing in Internet Protocols. Request for Comments 1102, Internet Engineering Task Force, May 1989.

[79] Matthew Caesar and Jennifer Rexford. BGP routing policies in ISP networks. *Network*, 19(6):5–11, November–December 2005.

[80] M. Steenstrup. An Architecture for Inter-Domain Policy Routing. Request for Comments 1478, Internet Engineering Task Force, June 1993.

[81] Martha E. Steenstrup. Inter-Domain Policy Routing Protocol Specification: Version 1. Request for Comments 1479, Internet Engineering Task Force, July 1993.

[82] G. Waters, J. Wheeler, A. Westerinen, L. Rafalow, and R. Moore. Policy Framework Architecture. Internet-draft, Internet Engineering Task Force, February 1999.

[83] CIM Policy Model White Paper, Version 1.1. White paper DSP0108, Distributed Management Task Force, May 2000.

[84] A. Westerinen, J. Schnizlein, J. Strassner, M. Scherling, B. Quinn, S. Herzog, A. Huynh, M. Carlson, J. Perry, and S. Waldbusser. Terminology for Policy-Based Management. Request for Comments 3198, Internet Engineering Task Force, November 2001.

[85] D. Durham, J. Boyle, R. Cohen, S. Herzog, R. Rajan, and A. Sastry. The COPS (Common Open Policy Service) Protocol. Request for Comments 2748, Internet Engineering Task Force, January 2000.

[86] R. Frye, D. Levi, S. Routhier, and B. Wijnen. Coexistence between Version 1, Version 2, and Version 3 of the Internet-standard Network Management Framework. Request for Comments 3584, Internet Engineering Task Force, August 2003.

[87] Dinesh C. Verma. Simplifying Network Administration Using Policy-Based Management. *Network*, 16(2):20–26, March–April 2002.

[88] J. L. van den Berg, R. Litjens, A. Eisenblätter, M. Amirijoo, O. Linnell, C. Blondia, T. Kürner, N. Scully, J. Oszmianski, and L. C. Schmelz. SOCRATES: Self-Optimisation and self-ConfiguRATion in wirelESs networks. In *4th Management Committee Meeting of the COST Action 2100*, February 2008.

[89] 3rd Generation Partnership Project (3GPP). Final Report on Self-Organisation and its Implications in Wireless Access Networks. Deliverable D5.9, January 2011.

[90] L. C. Schmelz, J. L. van den Berg, R. Litjens, K. Zetterberg, M. Amirijoo, K. Spaey, I. Balan, N. Scully, and S. Stefanski. Self-organisation in Wireless Networks – Use Cases and their Interrelation. In *22nd Meeting of the Wireless World Research Forum (WWRF)*. Wireless World Research Forum, May 2009.

[91] M. Amirijoo, L. Jorguseski, T. Kürner, R. Litjens, M. Neuland, L. C. Schmelz, and U. Türke. Cell Outage Management in LTE Networks. In *6th International Symposium on Wireless Communication Systems (ISWCS)*, pages 600–604. IEEE Communications Society, September 2009.

[92] Yi Zhang, Xiaoli Liu, and Weinong Wang. Policy Lifecycle Model for Systems Management. *IT Professional*, 7(2):50–54, March 2005.

[93] Gerd Wagner. Rule Modeling and Markup. In *1st International Conference on Reasoning Web*, pages 251–274. Springer LNCS, July 2005.

[94] OMG Object Constraint Language (OCL). Specification formal/2012-01-01, Object Management Group, January 2012.

[95] Information technology – Database languages – SQL. Standard ISO/IEC 9075:2011, International Organization for Standardization, December 2011.

[96] Alain Colmerauer and Philippe Roussel. The birth of Prolog. In *2nd Conference on History of Programming Languages (HOPL)*, pages 37–52. Association for Computing Machinery, April 1993.

[97] ILOG JRules. White paper, ILOG Corporation, October 2002.

[98] Drools Introduction and General User Guide. Documentation, The JBoss Drools team, May 2012.

[99] Friedman-Hill, Ernest. *Jess in Action: Java Rule-Based Systems*. Manning, October 2003.

[100] Jorge Lobo, Randeep Bhatia, and Shamim Naqvi. A Policy Description Language. In *16th National Conference on Artificial Intelligence (AAAI) and 11th Conference on Innovative Applications of Artificial Intelligence (IAAI)*, pages 291–298. Association for the Advancement of Artificial Intelligence, July 1999.

[101] APPEL – An Adaptable and Programmable Policy Environment and Language. Technical report CSM-161, University of Stirling, May 2011.

[102] Kevin Twidle, Naranker Dulay, Emil Lupu, and Morris Sloman. Ponder2: A Policy System for Autonomous Pervasive Environments. In *5th International Conference on Autonomic and Autonomous Systems (ICAS)*, pages 330–335. IEEE Computer Society, April 2009.

[103] OMG Unified Modeling Language (OMG UML), Infrastructure. Specification formal/2011-08-05, Object Management Group, August 2011.

[104] OMG Unified Modeling Language (OMG UML), Superstructure. Specification formal/2011-08-06, Object Management Group, August 2011.

[105] Nima Kaviani, Dragan Gašević, Milan Milanović, Marek Hatala, and Bardia Mohabbati. Model-Driven Engineering of a General Policy Modeling Language. In *9th International Workshop on Policies for Distributed Systems and Networks (POLICY)*, pages 101–104. IEEE Computer Society, June 2008.

[106] Common Information Model (CIM) Infrastructure. Standard DSP0004, Distributed Management Task Force, March 2010.

[107] CIM Policy Model White Paper, Version 2.7.0. White paper DSP0108, Distributed Management Task Force, June 2003.

[108] CIM Schema: Version 2.32.0. Standard, Distributed Management Task Force, 2011.

[109] John Strassner. DEN-ng: Achieving Business-Driven Network Management. In *8th Network Operations and Management Symposium (NOMS)*, pages 753–766. IEEE Communications Society, April 2002.

[110] John Strassner, José Neuman de Souza, David Raymer, Srini Samudrala, Steven Davy, and Keara Barrett. The Design of a New Policy Model to Support Ontology-Driven Reasoning for Autonomic Networking. *Networks and Systems Management*, 17(1-2):5–32, June 2009.

[111] John Strassner, Srini Samudrala, Greg Cox, Yan Liu, Michael Jiang, Jing Zhang, Sven van der Meer, Mícheál Ó Foghlú, and Willie Donnelly. The Design of a New Context-Aware Policy Model for Autonomic Networking. In *5th International Conference on Autonomic Computing (ICAC)*, pages 119–128. IEEE Computer Society, June 2008.

[112] Marshall Abrams and David Bailey. Abstraction and Refinement of Layered Security Policy. *Information Security: An Integrated Collection of Essays*, pages 126–136, June 1995.

[113] Jonathan D. Moffett. Policy Hierarchies for Distributed Systems Management. *Selected Areas in Communications*, 11(9):1404–1414, December 1993.

[114] René Wies. Using a Classification of Management Policies for Policy Specification and Policy Transformation. In *4th International Symposium on Integrated Network Management (IM)*, pages 44–56. IEEE Communications Society, May 1995.

[115] Barbara L. Dijker, editor. *A Guide to Developing Computing Policy Documents*. USENIX Association, 1996.

[116] Semantics of Business Vocabulary and Business Rules (SBVR), v1.0. Specification formal/2008-01-02, Object Management Group, January 2008.

[117] Steven Davy, Brendan Jennings, and John Strassner. The Policy Continuum – A Formal Model. In *2nd International Workshop on Modelling Autonomic Communications Environments (MACE)*, pages 65–79. IEEE Computer Society, October 2007.

[118] S. van der Meer, A. Davy, S. Davy, R. Carroll, B. Jennings, and J. Strassner. Autonomic Networking: Prototype Implementation of the Policy Continuum. In *1st International Workshop on Broadband Convergence Networks (BcN)*, pages 1–10, April 2006.

[119] J. Babiarz, K. Chan, and F. Baker. Configuration Guidelines for DiffServ Service Classes. Request for Comments 4594, Internet Engineering Task Force, August 2006.

[120] Thomas Koch, Christoph Krell, and Bernd Kraemer. Policy Definition Language for Automated Management of Distributed Systems. In *2nd International Workshop on Systems Management (SMW)*, pages 55–64. IEEE Computer Society, June 1996.

[121] Raouf Boutaba and Amine Benkiran. A Framework for distributed systems management. In *International Conference on Computer Networks, Architecture and Applications (NETWORKS)*, pages 287–298. North-Holland, October 1992.

[122] Raouf Boutaba and Simon Znaty. An Architectural Approach for Integrated Network and Systems Management. *Computer Communication Review*, 25(5):13–38, October 1995.

[123] Martín Abadi and Leslie Lamport. The Existence of Refinement Mappings. In *3rd Annual Symposium on Logic in Computer Science (LICS)*, pages 165–175. IEEE Computer Society, July 1988.

[124] Morris Sloman and Emil Lupu. Security and Management Policy Specification. *Network*, 16(2):10–19, March–April 2002.

[125] Taufiq Rochaeli and Claudia Eckert. Expertise Knowledge-Based Policy Refinement Process. In *8th International Workshop on Policies for Distributed Systems and Networks (POLICY)*, pages 61–65. IEEE Computer Society, June 2007.

[126] Kevin Carey and Vincent Wade. Using Automated Policy Refinement to Manage Adaptive Composite Services. In *11th Network Operations and Management Symposium (NOMS)*, pages 239–247. IEEE Communications Society, May 2008.

[127] Yathiraj B. Udupi, Akhil Sahai, and Sharad Singhal. A Classification-Based Approach to Policy Refinement. In *10th International Symposium on Integrated Network Management (IM)*, pages 785–788. IEEE Computer Society, May 2007.

[128] Arosha K. Bandara, Emil C. Lupu, Jonathan Moffett, and Alessandra Russo. A Goal-based Approach to Policy Refinement. In *5th International Workshop on Policies for Distributed Systems and Networks (POLICY)*, pages 229–239. IEEE Computer Society, June 2004.

[129] Arosha K. Bandara, Emil C. Lupu, Alessandra Russo, Naranker Dulay, Morris Sloman, Paris Flegkas, Marinos Charalambides, and George Pavlou. Policy Refinement for DiffServ Quality of Service Management. In *9th International Symposium on Integrated Network Management (IM)*, pages 469–482. IEEE Computer Society, May 2005.

[130] J. W. Lloyd. Practical Advantages of Declarative Programming. In *Joint Conference on Declarative Programming (GULP-PRODE), Volume 1*, pages 18–30. University of Valencia, September 1994.

[131] Dakshi Agrawal, James Giles, Kang-Won Lee, and Jorge Lobo. Policy Ratification. In *6th International Workshop on Policies for Distributed Systems and Networks (POLICY)*, pages 223–232. IEEE Computer Society, June 2005.

[132] Chetan Shankar and Roy Campbell. A Policy-based Management Framework for Pervasive Systems using Axiomatized Rule-Actions. In *4th International Symposium on Network Computing and Applications (NCA)*, pages 255–258. IEEE Computer Society, July 2005.

[133] Chetan Shiva Shankar, Anand Ranganathan, and Roy Campbell. An ECA-P Policy-based Framework for Managing Ubiquitous Computing Environments. In *2nd Annual International Conference on Mobile and Ubiquitous Systems: Networking and Services (MobiQuitous)*, pages 33–44. IEEE Computer Society, July 2005.

[134] E. Lupu and M. Sloman. Conflicts in Policy-Based Distributed Systems Management. *Software Engineering*, 25(6):852–869, November 1999.

[135] Jan Chomicki, Jorge Lobo, and Shamim Naqvi. Conflict Resolution Using Logic Programming. *Knowledge and Data Engineering*, 15(1):244–249, January–February 2003.

[136] Aashu Virmani, Jorge Lobo, and Madhur Kohli. Netmon: network management for the SARAS softswitch. In *7th Network Operations and Management Symposium (NOMS)*, pages 803–816. IEEE Communications Society, April 2000.

[137] Alfred Horn. On Sentences Which are True of Direct Unions of Algebras. *Symbolic Logic*, 16(1):14–21, March 1951.

[138] S. Gorton, C. Montangero, S. Reiff-Marganiec, and L. Semini. StPowla: SOA, Policies and Workflows. In *3rd International Workshop on Engineering Service-Oriented Applications: Analysis, Design and Composition (WESOA)*, pages 351–362. Springer LNCS, September 2007.

[139] Gavin A. Campbell and Kenneth J. Turner. Goals and Policies for Sensor Network Management. In *2nd International Conference on Sensor Technologies and Applications (SENSORCOMM)*, pages 354–359. IEEE Computer Society, August 2008.

[140] Carlo Montangero, Stephan Reiff-Marganiec, and Laura Semini. Logic-based detection of conflicts in APPEL policies. In *International Symposium on Fundamentals of Software Engineering (FSEN)*, pages 257–271. Springer LNCS, April 2007.

[141] Lynne Blair and Kenneth J. Turner. Handling Policy Conflicts in Call Control. In *8th International Conference on Feature Interactions in Telecommunications and Software Systems (ICFI)*, pages 39–57. IOS, June 2005.

[142] The ACCENT Policy Server. Technical report CSM-164, University of Stirling, July 2011.

[143] Kenneth J. Turner and Gavin A. Campbell. Goals and Conflicts in Telephony. In *10th International Conference on Feature Interactions in Software and Communications Systems (ICFI)*, pages 3–18. IOS, June 2009.

[144] N. Damianou, N. Dulay, E. Lupu, and M. Sloman. Tools for Domain-based Policy Management of Distributed Systems. In *8th Network Operations and Management Symposium (NOMS)*, pages 213–218. IEEE Communications Society, April 2002.

[145] Morris Sloman and Emil Lupu. Engineering Policy-Based Ubiquitous Systems. *Computer*, 53(7):1113–1127, September 2010.

[146] Sye Loong Keoh, Kevin Twidle, Nathaniel Pryce, Alberto E. Schaeffer-Filho, Emil Lupu, Naranker Dulay, Morris Sloman, Steven Heeps, Stephen Strowes, Joe Sventek, and Eleftheria Katsiri. Policy-based Management for Body-Sensor Networks. In *4th International Workshop on Wearable and Implantable Body Sensor Networks (BSN)*, pages 92–98. Springer IFMBE, March 2007.

[147] Alberto Schaeffer-Filho, Emil Lupu, Morris Sloman, Sye-Loong Keoh, Jorge Lobo, and Seraphin Calo. A Role-Based Infrastructure for the Management of Dynamic Communities. In *2nd International Conference on Autonomous Infrastructure, Management and Security (AIMS)*, pages 1–14. Springer LNCS, July 2008.

[148] Adele Goldberg and David Robson. *Smalltalk 80: The Language.* Addison-Wesley, 1989.

[149] Jonathan D. Moffett and Morris S. Sloman. The Representation of Policies as System Objects. In *Conference on Organizational Computing Systems (COCS)*, pages 171–184. Association for Computing Machinery, November 1991.

[150] Hang Zhao, Jorge Lobo, and Steven M. Bellovin. An Algebra for Integration and Analysis of Ponder2 Policies. In *9th International Workshop on Policies for Distributed Systems and Networks (POLICY)*, pages 74–77. IEEE Computer Society, June 2008.

[151] Robert Craven, Jorge Lobo, Jiefei Ma, Alessandra Russo, Emil Lupu, and Arosha Bandara. Expressive Policy Analysis with Enhanced System Dynamicity. In *4th International Symposium on Information, Computer, and Communications Security (ASIACCS)*, pages 239–250. Association for Computing Machinery, March 2009.

[152] Murray Shanahan. The Event Calculus Explained. In *Artificial Intelligence Today: Recent Trends and Developments*, pages 409–430. Springer, September 1999.

[153] Giovanni Russello, Changyu Dong, and Naranker Dulay. Authorisation and Conflict Resolution for Hierarchical Domains. In *8th International Workshop on Policies for Distributed Systems and Networks (POLICY)*, pages 201–210. IEEE Computer Society, June 2007.

[154] David Flater. Impact of Model-Driven Standards. In *35th Annual Hawaii International Conference on System Sciences (HICSS)*, page 285. IEEE Computer Society, January 2002.

[155] Kris Verlaenen, Bart De Win, and Wouter Joosen. Towards simplified specification of policies in different domains. In *10th International Symposium on Integrated Network Management (IM)*, pages 20–29. IEEE Communications Society, May 2007.

[156] Matthew Emerson and Janos Sztipanovits. Techniques for Metamodel Composition. In *6th OOPSLA Workshop on Domain Specific Modeling (DSM)*, pages 123–139. University of Jyväskylä, October 2006.

[157] Gerd Wagner, Grigoris Antoniou, Said Tabet, and Harold Boley. The Abstract Syntax of RuleML – Towards a General Web Rule Language Framework. In *International Conference on Web Intelligence (WI)*, pages 628–631. IEEE Computer Society, September 2004.

[158] Edgar F. Codd. Derivability, Redundancy and Consistency of Relations Stored in Large Data Banks. Report RJ 599, IBM Corporation, August 1969.

[159] E. F. Codd. A Relational Model of Data for Large Shared Data Banks. *Communications of the ACM*, 13(6):377–387, June 1970.

[160] Andrzej Uszok, Jeffrey M. Bradshaw, and Renia Jeffers. KAoS: A Policy and Domain Services Framework for Grid Computing and Semantic Web Services. In *2nd International Conference on Trust Management (iTrust)*, pages 16–26. Springer LNCS, March–April 2004.

[161] Lalana Kagal, Tim Finin, and Anupam Joshi. A Policy Language for a Pervasive Computing Environment. In *4th International Workshop on Policies for Distributed Systems and Networks (POLICY)*, pages 63–74. IEEE Computer Society, June 2003.

[162] Meta Object Facility (MOF) Core Specification. Specification formal/06-01-01, Object Management Group, January 2006.

[163] Extensible Markup Language (XML) 1.0 (Fifth Edition). Recommendation, World Wide Web Consortium, November 2008.

[164] Gerd Wagner, Adrian Giurca, and Sergey Lukichev. A Usable Interchange Format for Rich Syntax Rules Integrating OCL, RuleML and SWRL. In *Workshop on Reasoning on the Web (RoW)*, May 2006.

[165] J.-J. Ch. Meyer, F. P. M. Dignum, and R. J. Wieringa. The Paradoxes of Deontic Logic Revisited: A Computer Science Perspective. Technical report UU-CS-1994-38, Utrecht University, September 1994.

[166] B. Moore, E. Ellesson, J. Strassner, and A. Westerinen. Policy Core Information Model – Version 1 Specification. Request for Comments 3060, Internet Engineering Task Force, February 2001.

[167] NGOSS Architecture Technology Neutral Specification. Technical specification TMF053, TM Forum, May 2002.

[168] Information Framework (SID) Suite, Release 9.0. October GB922, TM Forum, May 2010.

[169] Krzysztof Czarnecki and Simon Helsen. Feature-Based Survey of Model Transformation Approaches. *Systems*, 45(3):621–645, July 2006.

[170] Tom Mens and Pieter Van Gorp. A Taxonomy of Model Transformation. In *International Workshop on Graph and Model Transformation (GraMoT)*, pages 125–142. Elsevier ENCTS, September 2005.

[171] Kevin Lano and David Clark. Model Transformation Specification and Verification. In *8th International Conference on Quality Software (QSIC)*, pages 45–54. IEEE Computer Society, August 2008.

[172] XSL Transformations (XSLT) Version 2.0. Recommendation, World Wide Web Consortium, January 2007.

[173] eXtensible Access Control Markup Language 2 (XACML) Version 2.0. Standard, Organization for the Advancement of Structured Information Standards, February 2005.

[174] John H. Gennari, Mark A. Musen, Ray W. Fergerson, William E. Grosso, Monica Crubézy, Henrik Eriksson, Natalya F. Noy, and Samson W. Tu. The Evolution of Protégé: An Environment for Knowledge-Based Systems Development. *Human-Computer Studies*, 58(1):89–123, January 2003.

[175] OWL 2 Web Ontology Language Document Overview. Recommendation, World Wide Web Consortium, October 2009.

[176] Web Services Description Language (WSDL) Version 2.0 Part 0: Primer. Recommendation, World Wide Web Consortium, June 2007.

[177] OWL-S: Semantic Markup for Web Services. Member submission, World Wide Web Consortium, November 2004.

[178] Siegfried Nolte. *QVT – Relations Language*. Springer, 2009.

[179] Siegfried Nolte. *QVT – Operational Mappings*. Springer, 2009.

[180] Tobias Bandh, Georg Carle, Henning Sanneck, and Lars-Christoph Schmelz. Automated Real Time Performance Management in Mobile Networks. In *1st WoWMoM Workshop on Autonomic Wireless AccesS (IWAS)*, pages 1–7. IEEE Computer Society, June 2007.

[181] LTE (Evolved UTRA) and LTE-Advanced radio technolgy. Technical specification 36 series, 3rd Generation Partnership Project (3GPP), September 2011.

[182] Next Generation Mobile Networks – Beyond HSPA & EVDO. White paper, NGMN Alliance, December 2006.

[183] Operations Requirements for the next generation of mobile networks. Draft working document OPRFQ-2009-01-1.4, TM Forum, May 2010.

[184] Telecommunication Management; Self-Organizing Networks (SON); Concepts and requirements (Release 11). Technical specification 32.500, 3rd Generation Partnership Project, December 2011.

[185] Self-Optimizing Networks: The Benefits of SON in LTE. White paper, 4G Americas, July 2011.

[186] Self Organizing Network – NEC's proposals for next-generation radio network management. White paper, NEC Corporation, February 2009.

[187] Self-Organizing Networks (SON) in 3GPP Long Term Evolution. White paper, Nomor Research, May 2008.

[188] Evolved Universal Terrestrial Radio Access Network (E-UTRAN); Self-configuring and self-optimizing network (SON) use cases and solutions (Release 9). Technical specification 36.902, 3rd Generation Partnership Project, April 2011.

[189] Javier Baliosian, Françoise Sailhan, Ann Devitt, and Anne-Marie Bosneag. The Omega Architecture: Towards Adaptable, Self-Managed Networks. In *1st International Workshop on Distributed Autonomous Network Management Systems (DANMS)*. University of Dublin, June 2006.

[190] T. Jansen, M. Amirijoo, U. Türke, L. Jorguseski, K. Zetterberg, R. Nascimento, L. C. Schmelz, J. Turk, and I. Balan. Embedding Multiple Self-Organisation Functionalities in Future Radio Access Networks. In *69th Vehicular Technology Conference (VTC Spring)*, pages 1–5. IEEE Vehicular Technology Society, April 2009.

[191] Lars-Christoph Schmelz, Mehdi Amirijoo, Andreas Eisenblaetter, Remco Litjens, Michaela Neuland, and John Turk. A Coordination Framework for Self-Organisation in LTE Networks. In *12th International Symposium on Integrated Network Management (IM)*, pages 193–200. IEEE Communications Society, May 2011.

[192] Evolved Universal Terrestrial Radio Access (E-UTRA) and Evolved Universal Terrestrial Radio Access Network (E-UTRAN); Overall Description; Stage 2 (Release 11). Technical specification 36.300, 3rd Generation Partnership Project, March 2012.

[193] Use Cases related to Self Organising Network, Overall Description. Deliverable, Next Generation Mobile Networks Alliance, May 2007.

[194] Tobias Bandh, Georg Carle, and Henning Sanneck. Graph Coloring Based Physical-Cell-ID Assignment for LTE Networks. In *5th International Wireless Communications and Mobile Computing Conference (IWCMC)*, pages 116–120. Association for Computing Machinery, June 2009.

[195] D. J. A. Welsh and M. B. Powell. An upper bound for the chromatic number of a graph and its application to timetabling problems. *Computer*, 10(1):85–86, May 1967.

[196] Mehdi Amirijoo, Pål Frenger, Fredrik Gunnarsson, Harald Kallin, Johan Moe, and Kristina Zetterberg. Neighbor Cell Relation List and Measured Cell Identity Management in LTE. In *11th Network Operations and Management Symposium (NOMS)*, pages 152–159. IEEE Communications Society, April 2008.

[197] Agreement on framework for PCI selection. Document R3-082357, 3rd Generation Partnership Project, August 2008.

[198] Nicholas Paine Gilman. *Profit Sharing Between Employer and Employee: A Study in the Evolution of the Wages System*. Houghton Mifflin Harcourt, 1893.

[199] Nizar Msadek. Optimierung eines SAP-basierten Bonusprozesses mit Web-UI-Komponenten für Beratungsunternehmen. Master's thesis, University of Augsburg, 2011.

[200] Frederick P. Brooks, Jr. No Silver Bullet – Essence and Accidents of Software Engineering. *Computer*, 20(4):10–19, April 1987.

List of Acronyms

3GPP	3rd Generation Partnership Project
ACI	Autonomic Computing Initiative
ACL	Access Control List
APPEL	Adaptable and Programmable Policy Environment and Language
AR	Administrative Region
ANSI	American National Standard for Information Technology
ACS	Autonomic Computing System
BGP	Border Gateway Protocol
BPMN	Business Process Model and Notation
CCO	Coverage and Capacity Optimization
CIM	Common Information Model
CIM	Computation-Independent Model
CM	Configuration Management
COPS	Common Open Policy Service
CPU	Central Processing Unit
DAC	Discretionary Access Control
DEN-ng	Directory Enabled Networks - next generation
DFG	German Research Foundation
DiffServ	Differentiated Services
DMTF	Distributed Management Task Force
DSL	Domain-Specific Language
EC	Event Calculus
ECA	Event Condition Action
ECAP	Event Condition Action Postcondition
ECPAP	Event Condition Precondition Action Postcondition
EGCI	EUTRAN Global Cell ID
ER	Entity Relationship
EUTRAN	Evolved UMTS Terrestrial Radio Access Network
FCAPS	Fault, Configuration, Accounting, Performance, and Security
FK	Foreign Key
FM	Fault Management
FSM	Finite State Machine
GSM	Global System for Mobile Communications
GPML	General Policy Modeling Language

GUI	Graphical User Interface
ICT	Information and Communication Technology
IDPR	Inter-Domain Policy Routing
IETF	Internet Engineering Task Force
IMEI	International Mobile Equipment Identity
IP	Internet Protocol
IPSec	IP Security
IPv4	Internet Protocol version 4
IT	Information Technology
KPI	Key Performance Indicator
LDAP	Lightweight Directory Access Protocol
LHS	Left Hand Side
LTE	Long Term Evolution
M2M	Model-to-Model
M2T	Model-to-Text
MAC	Mandatory Access Control
MDA	Model Driven Architecture
MDE	Model-Driven Engineering
MDSD	Model-Driven Software Development
MOF	Meta Object Facility
NE	Network Element
NGMN	Next Generation Mobile Networks
NGOSS	New Generation Operational Systems and Software
OAM	Operation, Administration, and Maintenance
OC	Organic Computing
OCL	Object Constraint Language
OCS	Organic Computer System
OMC	Operations and Maintenance Center
OMG	Object Management Group
OPEX	Operational Expenditure
OWL	Web Ontology Language
OWL-S	Semantic Markup for Web Services
PBM	Policy-Based Management
PBNM	Policy-Based Network Management
PCI	Physical Cell Identifier
PCIM	Policy Core Information Model
PDL	Policy Description Language
PDP	Policy Decision Point
PEP	Policy Enforcement Point
PIM	Platform-Independent Model
PK	Primary Key
PM	Performance Management
PSM	Platform-Specific Model
QoS	Quality of Service

QVT	Query/View/Transformation
R2ML	REWERSE I1 Rule Markup Language
RAT	Radio Access Technology
RBAC	Role-Based Access Control
RET	Remote Electrical Tilt
RFC	Request for Comments
RHS	Right Hand Side
SBVR	Semantics of Business Vocabulary and Business Rules
SID	Shared Information and Data Model
SLA	Service Level Agreement
SMC	Self-Managed Cell
SNMP	Simple Network Management Protocol
SOA	Service-Oriented Architecture
SON	Self-Organizing Network
ToS	Type of Service
TXP	Radio Transmission Power
UML	Unified Modeling Language
UMTS	Universal Mobile Telecommunications System
VoD	Video on Demand
VoIP	Voice over IP
WSDL	Web Services Description Language
XACML	eXtensible Access Control Markup Language
XML	Extensible Markup Language
XSL	Extensible Stylesheet Language
XSLT	XSL Transformation

List of Symbols

Ac	Set of modeled actions
ac, ac_i	Particular action
act, act_i	Particular active status of a policy
$Action$	Relation of actions
$Action_{la}$	View of the relation of actions $Action$ at layer la
AL	Relation of action links
AL_{la}	View of the relation of action links AL at layer la
Arg	Set of arguments
arg, arg_i	Particular argument
Be	Set of modeled binary expressions
be, be_i	Particular binary expression
BEL	Relation of binary expression links
BEL_{la}	View of the relation of binary expression links BEL at layer la
Cd	Set of modeled conditions
cd, cd_i	Particular condition
Co	Set of modeled concepts
co, co_i	Particular concept
$Concept$	Relation of concepts
$Concept_{la}$	View of the relation of concepts $Concept$ at layer la
$ConceptHierarchy$	Relation of concept hierarchy
$ConceptHierarchy_{la}$	View of the relation of concept hierarchy $ConceptHierarchy$ at layer la
$Condition$	Relation of conditions
$Condition_{la}$	View of the relation of conditions $Condition$ at layer la
$ConditionNesting$	Relation of condition nesting
$ConditionNesting_{la}$	View of the relation of condition nesting $ConditionNesting$ at layer la
DM	Domain model
DM_{la}	View of the domain model DM at layer la
EL	Relation of event links
EL_{la}	View of the relation of event links EL at layer la
Ev	Set of modeled events
ev, ev_i	Particular event
$Event$	Relation of events

$Event_{la}$	View of the relation of events $Event$ at layer la
Id_{Ac}	Set of identifiers for actions
Id_{Cd}	Set of identifiers for conditions
Id_{Co}	Set of identifiers for concepts
Id_{Ev}	Set of identifiers for events
Id_{La}	Set of identifiers for layers
Id_{OI}	Set of identifiers for operation invokings
Id_{Op}	Set of identifiers for operations
Id_{Pa}	Set of identifiers for parameters
Id_{Po}	Set of identifiers for policies
Id_{Pr}	Set of identifiers for properties
Id_{Re}	Set of identifiers for relationships
la, la_l	Particular layer
Lit	Set of possible literals
lit, lit_i	Particular literal
LM	Linking model
LM_{la}	View of the linking model LM at layer la
$MappingCo$	Relation of concept mappings
$MappingOp$	Relation of operation mappings
$MappingPa$	Relation of parameter mappings
$MappingPr$	Relation of property mappings
ml, ml_i	Particular mapping of literals
MM	Mapping model
mu, mu_i	Particular multiplicity
na, na_i	Particular navigability
no, no_i	Particular execution number of an action
o, o_i	Particular object
Oe	Set of modeled operational expressions
oe, oe_i	Particular operational expression
OEL	Relation of operational expression links
OEL_{la}	View of the relation of operational expression links OEL at layer la
oi, oi_i	Particular operation invoking
$OIL1$	First relation of operation invoking links
$OIL1_{la}$	View of the first relation of operation invoking links $OIL1$ at layer la
$OIL2$	Second relation of operation invoking links
$OIL2_{la}$	View of the second relation of operation invoking links $OIL2$ at layer la
$OIL3$	Third relation of operation invoking links
$OIL3_{la}$	View of the third relation of operation invoking links $OIL3$ at layer la
Op	Set of modeled operations
op, op_i	Particular operation

Operation	Relation of operations
Operation$_{la}$	View of the relation of operations *Operation* at layer *la*
Pa	Set of modeled parameters
pa, pa$_i$	Particular parameter
Parameter	Relation of parameters
Parameter$_{la}$	View of the relation of parameters *Parameter* at layer *la*
PM	Policy model
PM$_{la}$	View of the policy model *PM* at layer *la*
Po	Set of modeled policies
po, po$_i$	Particular policy
Policy	Relation of policies
Policy$_{la}$	View of the relation of policies *Policy* at layer *la*
Pr	Set of modeled properties
pr, pr$_i$	Particular property
Property	Relation of properties
Property$_{la}$	View of the relation of properties *Property* at layer *la*
Re	Set of modeled relationships
re, re$_i$	Particular relationship
ro, ro$_i$	Particular role
Relationship	Relation of relationships
Relationship$_{la}$	View of the relation of relationships *Relationship* at layer *la*
sc, sc$_i$	Particular concept
TraceAc	Relation of trace records for actions
TraceCd	Relation of trace records for conditions
TraceEv	Relation of trace records for events
TraceOI	Relation of trace records for operation invokings
TracePo	Relation of trace records for policies
ty, ty$_i$	Particular condition type
Type$_{Cd}$	Set of condition types
undef	Undefined value
Val, Val$_i$	Set of possible values
val, val$_i$	Particular value
Val$_B$	Set of possible boolean values
Val$_N$	Set of possible numeric values
Val$_T$	Set of possible textual values

List of Figures

List of Tables

List of Algorithms

Curriculum Vitae

Professional Experience

February 2006 to October 2012:	**Research Associate** Software Methodologies for Distributed Systems, University of Augsburg, Germany
November 2004 to March 2005:	**Intern** Fujitsu Siemens Computers Ltd., Bracknell, UK
April 2003 to November 2004:	**Freelancer** Institute of Computer Science, University of Augsburg, Germany
April 2001 to November 2004:	**Freelancer** buecher.de GmbH & Co. KG, Augsburg, Germany
September 2000 to October 2000:	**Intern** Verlagsgruppe Weltbild GmbH, Augsburg, Germany

Education

October 2000 to January 2006:	**Applied Computer Science with an Economics Minor (Diploma)** University of Augsburg, Germany
August 1999 to July 2000:	**Civilian Service** Nachsorge Zentrum Augsburg, Germany
September 1990 to June 1999:	**Secondary School** Rudolf-Diesel-Gymnasium Augsburg, Germany
September 1986 to July 1990:	**Primary School** Grundschule Augsburg Hochzoll-Süd, Germany